AF846587

City Development: Issues and Best Practices

Editors-in-Chief

Huhua Cao, Geography, Environment and Geomatics, University of Ottawa, Ottawa, Canada

John Zacharias, College of Architecture and Landscape, Peking University, Beijing, China

Claude Ngomsi, United Nations Human Settlements Programme (UN-Habitat), Nairobi, Kenya

The current rate of urbanization is unprecedented and poses enormous challenges for governing bodies. New approaches to development and urban management are needed in the context of globalization and the need for local sustainability. While the developing world itself offers an abundance of lessons, case studies and best practices, these have rarely been positioned as cutting-edge contributions to reformed practices in city development. It is well recognized that the experience of the developed world is an incomplete guide to the new challenges posed by urbanization in the contemporary world.

The "City Development: Issues and Best Practices" book series includes academic research, comparative and applied research, and case studies at the scale of the neighborhood, city, region, nation and supranational levels. This series will offer an opportunity to present the latest academic research and best practices in urban development with the goal of promoting sustainable and inclusive development, learning from the diverse and complementary experiences of rapidly urbanizing areas of the world. Although this book series focuses primarily on the developing world, we intend to include the latest academic research and evolving best practices from developed countries.

The series is intended for geographers, planners, engineers, urban designers, architects, political scientists, sociologists, and economists, as well as policy makers and representatives from government, civil society, industry, etc. who are interested in the developing world. The series will also interest students seeking a foundation in the comparative analysis of key issues facing rapidly urbanizing areas of the world. It will include monographs, edited volumes and textbooks. Book proposals and final manuscripts will be peer-reviewed.

The areas to be covered in the series include, but are not limited to, the following:

- Urban Equity and Inclusivity
- Reforming Informal Settlements
- Climate Change and Adaptation of the Built Environment
- Health Crises and Urban Risk Management
- Privacy, Surveillance, Security and Collective Wellbeing
- Governance and Urban Resilience
- Participatory Policy and the Right to the City
- Indigenous and Ethnic Minority Urbanization
- Mobility and Urban Transformation
- Land Use and Urban Landscapes
- Urban Infrastructure and Transport Systems
- Urban Intelligence and Technologies
- Participatory Budgeting and Community Building
- Industrial Parks and Agro-Processing Zones
- Housing and Land Tenure Issues

Ria Perrault · Jennifer Chinweokwu Ogborogu ·
Rebecca Nagar · Huhua Cao

Sub-Saharan African Sustainable Development

Climate Change, Energy, and Urbanization

Springer

Ria Perrault
Department of Geography, Environment and Geomatics
University of Ottawa
Ottawa, ON, Canada

Jennifer Chinweokwu Ogborogu
Department of Geography, Environment and Geomatics
University of Ottawa
Ottawa, ON, Canada

Rebecca Nagar
Department of Geography, Environment and Geomatics
University of Ottawa
Ottawa, ON, Canada

Huhua Cao
Department of Geography, Environment and Geomatics
University of Ottawa
Ottawa, ON, Canada

ISSN 2731-7773 ISSN 2731-7781 (electronic)
City Development: Issues and Best Practices
ISBN 978-981-95-1457-1 ISBN 978-981-95-1458-8 (eBook)
https://doi.org/10.1007/978-981-95-1458-8

© The Editor(s) (if applicable) and The Author(s), under exclusive license to Springer Nature Singapore Pte Ltd. 2025

This work is subject to copyright. All rights are solely and exclusively licensed by the Publisher, whether the whole or part of the material is concerned, specifically the rights of translation, reprinting, reuse of illustrations, recitation, broadcasting, reproduction on microfilms or in any other physical way, and transmission or information storage and retrieval, electronic adaptation, computer software, or by similar or dissimilar methodology now known or hereafter developed.
The use of general descriptive names, registered names, trademarks, service marks, etc. in this publication does not imply, even in the absence of a specific statement, that such names are exempt from the relevant protective laws and regulations and therefore free for general use.
The publisher, the authors and the editors are safe to assume that the advice and information in this book are believed to be true and accurate at the date of publication. Neither the publisher nor the authors or the editors give a warranty, expressed or implied, with respect to the material contained herein or for any errors or omissions that may have been made. The publisher remains neutral with regard to jurisdictional claims in published maps and institutional affiliations.

This Springer imprint is published by the registered company Springer Nature Singapore Pte Ltd.
The registered company address is: 152 Beach Road, #21-01/04 Gateway East, Singapore 189721, Singapore

If disposing of this product, please recycle the paper.

Acknowledgements

The creation of this book has been a collaborative endeavor, drawing on the expertise and insights of scholars, practitioners, and community leaders. We extend our deepest gratitude to all contributors who shared their knowledge and experiences.

Special thanks go to Jack Cao, Ebenezer Kumi, and Kalan Watanabe Watanabe, whose dedicated research and thoughtful analyses form the backbone of this book. We would also like to express our sincere appreciation to Dr. Navid Mashhadi Moghaddam for his invaluable contributions in finalizing this manuscript.

Introduction

Sub-Saharan Africa is a region teeming with diversity and potential, characterized by a rich mosaic of cultures, languages, and ecosystems. This is a part of the world where rapid urbanization coexists with deeply rooted traditions, where economic ambitions are tempered by environmental constraints, and where the challenges of the twenty-first century call for innovative and sustainable solutions. In this context, the book: *Sub-Saharan African Sustainable Development* aims to explore the intricate dimensions of sustainability, offering a comprehensive examination of the urgent issues facing the region while proposing potential pathways toward a sustainable future.

The Imperative of Sustainable Development in Sub-Saharan Africa

The concept of sustainable development—ensuring that the needs of the present are met without jeopardizing the ability of future generations to meet their own—is particularly pertinent in Sub-Saharan Africa. The region faces a host of unique challenges that make this goal both urgent and complex. One of the most pressing issues is the rapid urbanization that is transforming cities across the region. Urban centers are expanding at an unprecedented pace, often outstripping the development of infrastructure and services needed to support such growth. This phenomenon places immense pressure on resources, exacerbates environmental degradation, and magnifies socio-economic inequalities. In addition to urbanization, Sub-Saharan Africa is highly vulnerable to the impacts of climate change. Extreme weather events, shifting agricultural zones, and growing water scarcity threaten not only the livelihoods of millions but also the broader stability of food systems and economies. These challenges are compounded by limited access to reliable and affordable energy. While the region is endowed with significant renewable energy resources, the task of harnessing these sustainably and equitably remains daunting. Environmental degradation is another critical concern. Deforestation, pollution, and the loss of biodiversity are often driven by unsustainable resource extraction and poor land-use practices. These issues are further complicated by deep-seated social inequities. Poverty, gender inequality, and social exclusion persist, undermining efforts to achieve inclusive and equitable development.

The interconnected nature of these challenges calls for holistic solutions that balance economic growth, environmental stewardship, and social equity. It is within this context that the need for sustainable development in Sub-Saharan Africa becomes not only an opportunity but an imperative.

Purpose and Structure of the Book

This book is organized into three distinct parts, each addressing a crucial aspect of sustainable development in the region. The first part explores the role of urban green infrastructure in managing thermal conditions within rapidly growing cities, with a particular focus on Sub-Saharan Africa's informal settlements. The second shifts attention to renewable energy development, using Cameroon as a case study to illustrate broader regional dynamics. The final part delves into the urbanization challenges in Nigeria, offering insights that have relevance across the entire Sub-Saharan region.

By structuring the book around these themes, we aim to offer a blend of theoretical frameworks, real-world case studies, and practical applications that illuminate the pathways toward sustainability in Sub-Saharan Africa.

Part 1: Urban Green Infrastructure and Thermal Regulation
In the first part, the book addresses the vital role that urban green infrastructure can play in managing thermal conditions in Sub-Saharan Africa's cities that are experiencing rapid growth. In many of these cities, informal settlements—often lacking basic services and adequate infrastructure—are disproportionately affected by extreme heat. These areas, with their dense populations and lack of greenery, are particularly vulnerable to the health and well-being impacts of urban heat islands. The discussions begin by contextualizing the challenges faced by informal settlements, highlighting how inadequate urban planning has left many of these areas exposed to harsh environmental conditions. From here, the book introduces key theoretical frameworks from urban ecology and climate adaptation, emphasizing how green infrastructure—such as parks, green roofs, and urban forests—can serve as natural regulators of temperature. The importance of thermal regulation cannot be overstated. Prolonged exposure to extreme heat has serious health implications, particularly for vulnerable populations. This section delves into how integrating natural elements into urban planning not only enhances the livability of cities but also strengthens their resilience in the face of climate change.

Part 2: Renewable Energy Development and Resilience
The second part of the book shifts focus to renewable energy, examining its potential to catalyze sustainable development across the region. Using Cameroon as a focal point, this section provides an in-depth look at the region's renewable energy resources, including solar, wind, hydro, and biomass. However, the development of renewable energy is not just a technical challenge—it is also a social and ecological one. The book explores socio-technical and socio-ecological perspectives, emphasizing the need to consider how technological choices intersect with social systems and ecological constraints. Energy development in Sub-Saharan Africa cannot succeed without addressing these intertwined factors. A key theme in this section is energy governance. The book analyzes the policy frameworks, institutional capacities, and governance challenges that shape energy development in the region. It also examines the broader impacts of energy projects on critical resources such as water and food security, as well as community health and well-being. Through this lens, renewable energy emerges as a powerful tool for sustainable development—one that, if pursued with careful attention to equity and ecological dynamics, can address many of the region's most pressing needs.

Part 3: Urban Development and Sustainability
The final part delves into the specific case of Nigeria, whose rapid urbanization mirrors broader trends across Sub-Saharan Africa. The discussion begins by tracing the historical factors that have driven Nigeria's urban growth, from rural–urban migration to economic shifts that have concentrated populations in major cities. The environmental and social impacts of this unplanned urban expansion are severe. Deforestation, pollution, and the displacement of communities are just some of the consequences of unchecked growth. These challenges are not unique to Nigeria but resonate across the region, making this section's insights highly relevant to a broader audience. In response to these challenges, the book presents strategies for sustainable urban development. Policy approaches, governance reforms, and community-based interventions are discussed as potential solutions to the urban sustainability crisis. Case studies of successful initiatives are used to demonstrate the transformative potential of applying sustainability principles to urban development. This part offers a nuanced understanding of Nigeria's urbanization challenges while showcasing practical solutions that can inform broader regional efforts. The lessons learned from Nigeria's experience hold valuable insights for any part of Sub-Saharan Africa grappling with the complexities of urban growth and sustainability.

Overall, this book seeks to provide a comprehensive understanding of the sustainability challenges facing Sub-Saharan Africa, while offering practical pathways forward. By addressing the intricate balance between urbanization, energy development, and environmental protection, the book serves as both a critical analysis and a call to action for sustainable development in the region.

Theoretical Underpinnings and Key Concepts

Throughout this book, several theoretical frameworks serve as the foundation for our discussions, offering a coherent lens through which the complexities of sustainable development in Sub-Saharan Africa can be understood.

First and foremost is the principle of *sustainable development*, which guides our exploration of how to balance economic growth with environmental protection and social equity. This overarching concept informs every discussion, framing sustainability as a dynamic interplay between competing needs. Next is *political ecology*, a critical framework that examines the power dynamics and governance structures shaping environmental outcomes and resource distribution. By applying this lens, we can better understand the influence of politics, economics, and social hierarchies on sustainability efforts. *Socio-technical systems* theory provides another important perspective, focusing on how social factors and technological systems co-evolve and influence sustainable development trajectories. This concept highlights the need for an integrated approach, recognizing that technological advancements alone cannot drive progress without considering the social context in which they operate. Finally, the idea of *socio-ecological resilience* underscores the capacity of communities and ecosystems to withstand and adapt to environmental stresses and shocks. As Sub-Saharan Africa faces increasing environmental challenges, understanding how resilience can be fostered at both community and ecosystem levels is crucial.

By grounding the discussions in these theoretical frameworks, the book offers depth and coherence to the multifaceted issues at hand, providing readers with a structured approach to understanding sustainability in this region.

The Importance and Urgency of the Topic

The stakes for sustainable development in Sub-Saharan Africa are immense. The decisions made today will determine the region's environmental health, economic prosperity, and social stability for generations. The urgency of this issue is amplified by several critical factors.

One of the most pressing is the acceleration of climate change. Sub-Saharan Africa is already experiencing the severe impacts of the global climate crisis, from shifting agricultural zones to extreme weather events. The region's vulnerabilities demand immediate and decisive action to mitigate these effects and adapt to the evolving climate. In parallel, demographic trends present both challenges and opportunities. The region's burgeoning youth population represents a potential demographic dividend, but harnessing this opportunity requires creating sustainable pathways for education, employment, and civic engagement. Failure to do so could lead to widespread economic instability and social unrest. Moreover, the region's development trajectory has implications far beyond its borders. In our increasingly interconnected world, Sub-Saharan Africa's efforts to achieve sustainability are intrinsically linked to global initiatives aimed at addressing climate change, biodiversity loss, and the Sustainable Development Goals (SDGs). What happens in this region will ripple outward, affecting global progress on these critical issues.

This book not only seeks to analyze the pressing challenges but also aims to inspire action by showcasing viable solutions and encouraging collaboration across sectors. It is a call to collective responsibility—one that underscores the interconnectedness of the local and the global, and the shared fate we all have in shaping a sustainable future.

Setting the Stage for Detailed Discussions

The chapters of this book are designed to build upon each other, creating a narrative that captures the complexity and diversity of sustainable development in Sub-Saharan Africa. Each part provides a deeper dive into the challenges and opportunities specific to the region, while highlighting overarching themes that run throughout the discussions.

As we move forward, the book strives to illuminate the intricate interactions between environmental, social, and technological factors. The aim is not merely to present problems but to explore the interplay between these factors, demonstrating the complexity of sustainable development and the need for integrated, multi-faceted approaches. At the same time, the book offers practical insights, bringing theoretical concepts to life through real-world case studies and examples. These examples are drawn from across the region, illustrating how communities, policymakers, and practitioners are applying sustainability principles in diverse contexts. Crucially, the book encourages critical thinking. It invites readers to question assumptions, engage with alternative perspectives, and think deeply about the material presented. Sustainability is not a one-size-fits-all proposition, and this book emphasizes the importance of tailoring solutions to fit the unique social, environmental, and economic realities of each community and country.

Beyond analysis, the book aims to foster hope and motivation. By highlighting successful initiatives and potential pathways forward, it offers a sense of optimism and agency, encouraging readers to see themselves as active participants in the journey toward a sustainable future.

A Call for Tailored Solutions and Collaborative Action

As we embark on this exploration of sustainable development in Sub-Saharan Africa, it is important to acknowledge that there are no universal solutions. The region's immense diversity demands approaches that are sensitive to local contexts. What works in one country or community may not work in another. However, despite this diversity, certain common threads emerge—such as the need for inclusive governance, community engagement, and integrated planning—all of which are critical components of any effective sustainable development strategy.

This book is an invitation to policymakers, scholars, practitioners, and community members to engage with the challenges and opportunities that lie ahead. The potential to steer Sub-Saharan Africa toward a sustainable and prosperous future is vast, but it requires collaborative effort. By embracing innovative solutions grounded in sound theoretical frameworks and tailored to local realities, the region can navigate its complex development challenges and move toward a more sustainable path.

As we delve into the detailed discussions that follow, let us proceed with a spirit of inquiry, openness, and determination. The future of Sub-Saharan Africa—and, indeed, the world—depends on our ability to engage with these challenges and act with purpose and foresight.

Contents

Part I
Urban Green Infrastructure and Thermal Regulation

Chapter 1
Introduction to Urban Green Infrastructure

1.1 Introduction

In sub-Saharan African cities, two key trends are on the rise: climate change leading to increasing temperatures, and rapid population growth. Natural population growth as well as migration from rural areas to major cities throughout the continent has led to Africa being home to 79 of the 100 fastest-growing cities in the world (Global Center on Adaptation 2022). Urban sub-Saharan Africa (SSA) faces several unique challenges along with high and increasing populations, such as poverty and informal settlements and basic urban infrastructure, the provision of which has not kept up with the rate of population growth.

Within SSA cities, climate change is already having substantial impacts on human health and livelihoods. Among these impacts, extreme heat and excessively high temperatures has among the most significant tolls on human health, linked to increased risks of mortality and morbidity, and made worse by the Urban Heat Island (UHI) effect within cities (Jagarnath et al. 2020; Mukherji et al. 2023). The UHI is defined as "the phenomenon of higher urban temperatures than the rural surroundings and affects temperatures in the air and at [land surfaces]" (Li et al. 2022a: 4). It amplifies heat: when natural surfaces such as vegetated areas are replaced by impervious ones, surface albedo is altered and radiation is trapped within street canyons, all while anthropogenic activities generate additional heat within the city (Li et al. 2022b; Van de Walle et al. 2022). It is well-established that UHI amplifies warming, but in climate model analyses it is difficult to observe and predict, especially in terms of human impacts (Parkes et al. 2022).

Moreover, the human impacts of the UHI are not equally distributed. Wealthy groups and neighbourhoods have access to a wider range of adaptation options for extreme heat, such as air conditioning, and often benefit from the ecosystem services of well-established and healthy trees and green spaces near their homes. Meanwhile the urban poor, including residents of informal settlements (including slums), have access to a smaller array of potential adaptive responses and inhabit morphological

© The Author(s), under exclusive license to Springer Nature Singapore Pte Ltd. 2025
R. Perrault et al., *Sub-Saharan African Sustainable Development*, City Development: Issues and Best Practices, https://doi.org/10.1007/978-981-95-1458-8_1

types that tend to further amplify the UHI. As such, the potential human impacts of temperature extremes are likely to be far greater to those of lower socioeconomic status, including inhabitants of informal settlements, approximately 60% of SSA's urban residents (Mukherji et al. 2023).

Nonetheless, the Global Center on Adaptation (GCA) reported that since many African cities are still in relatively early stages of urbanization with much growth still to come, "there is a unique opportunity to get things right" (Global Center on Adaptation 2022: 17). By recognizing threats and potential vulnerabilities at early stages of development, cost-effective 'no regrets' adaptation measures can be leveraged to have immense impacts for decades to come.

Using nature and green space as a climate solution is one such measure, identified by the GCA, African Union, United Nations International Panel on Climate Change (IPCC), and state governments alike. Nature-based solutions (NBS) is identified as a priority intervention area in the African Union Climate Change and Resilient Development Strategy and Action Plan (2022–2032), while over a third of the priority adaptation actions determined by African countries' Nationally Determined Contributions under the Paris Agreement involve nature-based solutions or adaptation measures. One type of NBS is urban green infrastructure (UGI).[1]

NBS, UGI, and other related concepts have been growing in popularity in recent years, but the existing scholarship is mostly oriented towards Western contexts and its applications in the Global North. Research gaps have been identified on the applicability, feasibility, governance, utility, and practical implementation of UGI and associated ecosystem services for African cities (Adegun 2019; du Toit et al. 2018; Lindley et al. 2018; Nassary et al. 2022; Wijesinghe and Thorn 2021). Heat waves, UHI, and extreme heat are under-researched among climate events, perhaps due to their transient nature, leaving little physical evidence of their severity (in contrast to more dramatic and acute climate impacts such as flooding), despite the human and health impacts (Laue et al. 2022). The adaptive responses of urban residents and systems to extreme temperatures, as well as the vulnerabilities of particular residents and groups within SSA cities appears to be a gap in the peer-reviewed literature as well (Hunter et al. 2020).

The vulnerability of human health and our systems of support is now clearer than ever. Throughout the COVID-19 pandemic, socioeconomically marginalized and underserved communities worldwide faced the highest risk burden, with populations of low and middle-income countries facing increased mortality from COVID-19, and economic status also potentially acting as a risk factor, although this is not always the case (Hassan et al. 2020). So too does socioeconomic status play a role in the distribution of vulnerability to risks associated with heat exposure. This plays out at the national level, with low and middle-income countries who have contributed the least to the causes of climate change bearing a disproportionate amount of its burden; but it also plays out at the community, household, and individual level (Mukherji et al.

[1] Understood in this work as "all natural, semi-natural and artificial networks of multifunctional ecological systems within, around and between urban areas, at all spatial scales" (Tzoulas et al. 2007, as quoted by du Toit et al. 2018: 1).

2023). This dynamic can be understood through the lens of urban political ecology (UPE), a theoretical framework that is more frequently applied to climate hazards such as flooding in the developing world.

This research aims to investigate the UHI in SSA cities through an urban political ecology perspective, to understand the current vulnerabilities of SSA cities with respect to the spatial distribution of heat relative to informal settlements, and the potential role of urban green infrastructure. The UPE theoretical lens will help show how the socioeconomic relations between societal groups is reflected in cities and how socio-spatial arrangements interact with and a changing climate to contribute to vulnerability.

This work will begin by introducing the theoretical framework, followed by an explanation of the social and environmental contexts of extreme heat in SSA cities. The challenges of increasing temperatures in urban environments, along with common solutions and responses are then outlined. Subsequently, a discussion of urban green infrastructure with respect to ecosystem services, informal settlements, and dimensions of vulnerability is offered. Finally, the theoretical framework is revisited, to discuss the economic, ecological, and political influences on vulnerability to heat stress, role of green infrastructure, with implications for environmental justice and future research.

1.2 Theoretical Framework and Key Concepts

1.2.1 Urban Political Ecology (UPE)

UPE is a field of research and conceptual approach that takes the perspective that urbanization is a political, social, and ecological process wherein cities are not outside of nature but a type of it. The field of UPE is decidedly political, viewing human-nature interactions as being shaped by socioeconomic forces and asking questions about who shapes socio-natural and socio-spatial configurations, and who is served by this exercise of power (Heynen et al. 2005). Under this understanding, landscapes and urban forms are the product of humanity's interactions with the natural world (Marks and Connell 2023).

These interactions are also argued to both reflect and reproduce existing societal power dynamics, viewing urbanization as "the transformation of nature through processes of capital accumulation" (Silver 2015: 986). Urban metabolism, which is understood as the series of interconnected, dynamic, and contested processes between humans and non-humans that continuously quantitatively and qualitatively transforms and rearranges these groups, brings these concepts of transformation of nature and societal power together (Swyngedouw 2006). The relationships between the environment and socioeconomic and political contexts within cities is reflected in the relationship with nature (the socio-natural relationship) as well as spatial relationships between different human groups, and their surroundings (the socio-natural relationship) (Keil 2003). In this way, UPE draws attention to how certain areas and groups are unequally exposed to natural and unnatural hazards and vulnerability (Salami et al. 2018).

1.2.2 Vulnerability

There are many approaches to viewing vulnerability. Busby et al. (2013) outline several potential approaches to understanding vulnerability, thereby highlighting its conceptual fragmentation. However, in Adger's (2006) foundational work on the subject, the author identifies three common parameters of vulnerability present in all formulations of the concept: a system's exposure to stress, sensitivity, and adaptive capacity. In line with this, the IPCC's fourth assessment report offers a widely-used, climate-specific definition of vulnerability, as follows:

> "Vulnerability is the degree to which a system is susceptible to, and unable to cope with, adverse effects of climate change, including climate variability and extremes. Vulnerability is a function of the character, magnitude, and rate of climate change and variation to which a system is exposed, its sensitivity, and its adaptive capacity." (IPCC 2007: 21)

Although this definition explicitly applies to climate change, it aligns with Adger's view while also encompassing the interactions between physical risk or exposure and social susceptibility to harm. In this way, the concept of vulnerability can serve as an important analytical tool for guiding normative analysis (Adger 2006).

1.2.3 Adaptive Capacity

> "Adaptive capacity is the ability of a system to adjust to climate change (including climate variability and extremes) to moderate potential damages, to take advantage of opportunities, or to cope with the consequences." (IPCC 2007: 21)

Adaptive capacity is closely linked to vulnerability, particularly in the context of climate change. It is a combination of physiological, technical, and economic options, broadly determined by state capacity and level of development, as well as individual wealth (Parkes et al. 2022). Societies with low levels of development and generally weak state capacity will have limited capacity to adapt to climate change and vice versa for societies with higher levels of development and state capacity. At the individual level, richer people have access to more potential options for modifying their environment in the face of risk than poorer people, for example, by purchasing and installing an air conditioner in response to a heat wave. Poverty and relative poverty limits adaptive capacity (Serdeczny et al. 2017).

Unfortunately, half of the people in the world who could be classified as extremely poor as defined by the World Bank (earning less than $1.90USD per day) are living in Africa, approximately 490 million people as of 2018 (Welborn 2018). A higher adaptive capacity enables communities to respond to climate change impacts and hazards more effectively, while lower adaptive capacity-communities are less able to adapt, prepare, and respond to various increased risks. Low- and middle-income nations with large populations of informal settlement dwellers in Africa generally have the cities with lowest adaptive capacity, as these factors are frequently combined with infrastructure and service provision deficits (Hunter et al. 2020). Multidimensional poverty, poor governments, and political instability limits many African states' ability to respond to the demands of a changing climate, and address current and future vulnerabilities (Ncongwane et al. 2021).

1.3 The Social Context of Extreme Heat: Population and Urbanization

Africa's population is rapidly increasing. In the year 2000, the continent's population was 0.82 billion; in 2022, it was 1.43 billion and this number is projected to reach 2.54 billion by 2050 (Parkes et al. 2022). Amidst the growth in population, the continent's population distribution is adjusting in turn. In 1960, only 15% of sub-Saharan Africa's population resided in urban areas; this number doubled by 1996, and continues to increase steadily, sitting at 42% by 2023 according to data from the World Bank (United Nations Population Division 2023). As a result of the continent's overall population growth, and the trend of rural to urban migration, Africa's urban population has increased by over 2000% since 1950, while the number of urban agglomerations reached 7617 in 2015 (up from 624 agglomerations 65 years prior) (OECD/SWAC 2020) (Fig. 1.1).

Because of the significant population growth that Africa is experiencing it is increasingly important that the development African cities undergo is sustainable (Cavan et al. 2014; Titz and Chiotha 2019). Successfully managing urban growth has been a challenge for SSA cities, which exhibit high spontaneity in their patterns of development. However, the manner in which growth occurs and the city is subsequently formed can be a key driver of climate impacts and in turn, the actualized impacts on urban populations (Worku 2017).

Urban population growth, in combination with the strength of urban management, can impact on the health of urban dwellers, especially in the context of excessive heat (Smit and Parnell 2012). There is a synergistic interaction between population growth and dangerous heat that could have significant impacts on future urban thermal environments and livability due to the rate of heat stress events (Li et al. 2022a). Population changes, and population density, have an important influence on the number of potential heat stress events possible in a given region, where 'heat stress event' is a function of apparent temperature and discomfort index[2] (Parkes et al. 2022). Figure 1.2 shows how the total global exposure to extreme heat has increased since the 1980s. While both population growth and urban warming are

[2] Refer to Parkes et al. 2022 for further explanation.

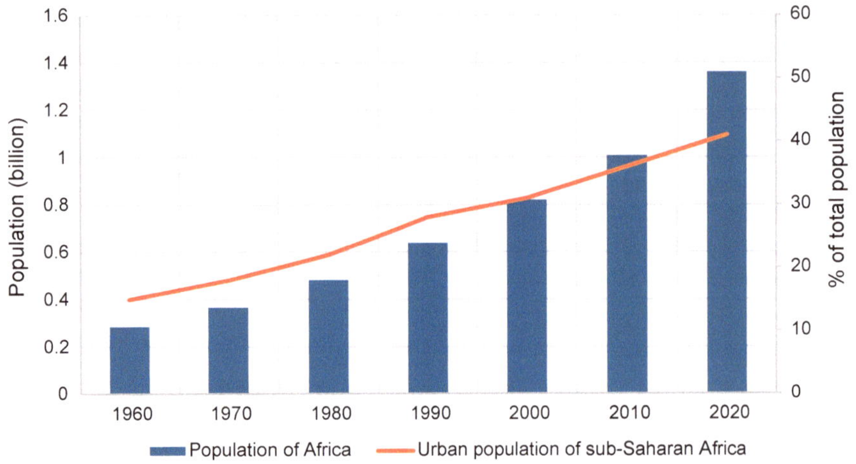

Fig. 1.1 Population of Africa and share of urban population for sub-Saharan Africa between 1960–2020. Data from Sources: Roser and Rodés-Guirao (2013) (revised 2021) and The World Bank (2023)

contributing to increased heat exposure, the steeper rate of change of population indicates that growing populations are a critical driver of urban exposure to heat.

With increasing urban populations, evidence of anthropogenic land use and land cover change (LULCC) through urbanization is rising as well. Land use, or human activities on land, and land cover, the biophysical characteristics of the land's surface, directly impact one another. A particular type of land cover could enable certain land uses, but not others. Anthropogenic land use, such as paving and building to support growing demand for civic infrastructure in urbanizing areas alters the land cover. Urbanization tends to modify the thermal characteristics of the land, and acts as a direct driver of urban thermal dynamics and the subsequent UHI. Other factors include seasonal variations, building materials, urban morphology, and factors such as altitude and vegetation density (Li et al. 2022a; Orimoloye et al. 2019).

Vegetation and urban green spaces are under threat globally from urbanization and urban expansion. This is true in Africa as well, where bushland, grassland, and crops are frequently converted into bare land for residential or industrial development, and green spaces are being fragmented and disappearing in favour of development and resource consumption (Wijesinghe and Thorn 2021). However, maintaining green space, and access to it, is important not only for the health of residents, but the well-being of cities themselves. Vegetated landscapes and green spaces have a positive impact on urban energy exchanges, regulating temperatures and providing widely-recognized benefits including flood regulation, ground stabilization, and heat stress relief (Cavan et al. 2014). Especially in developing countries, population and economic expansion have primarily driven LULCC through urbanization.

Rural–urban migration is reflected in cities' unregulated growth patterns, which has interacted with poverty to result in an increasing number of informal settlement

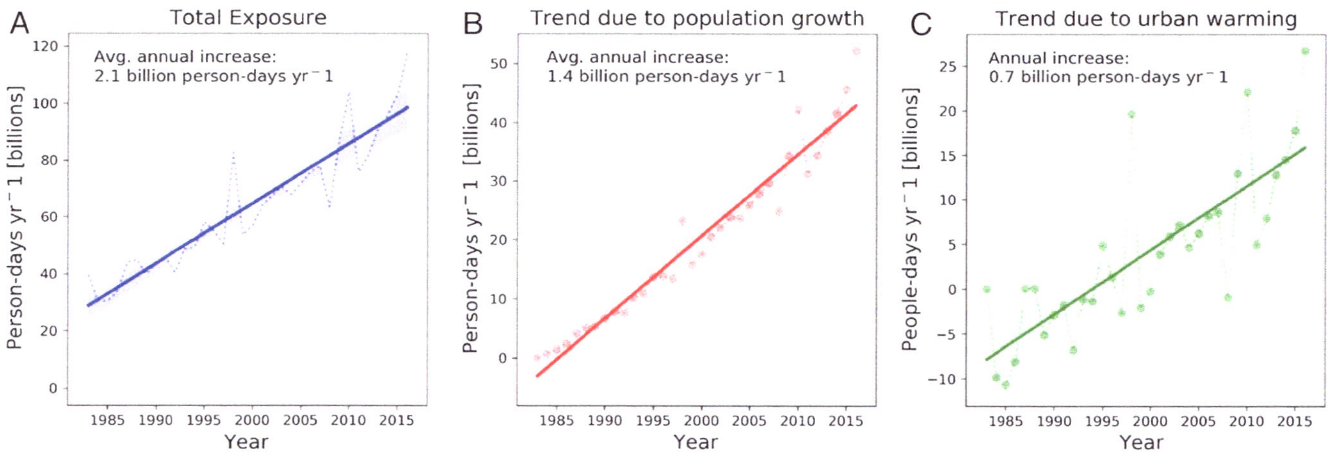

Fig. 1.2 Global urban population exposure to extreme heat from 1983 to 2016 (**A**), with contribution from population growth (**B**), and total urban warming (**C**) decoupled (Tuholske et al. 2021)[3]

[3] Extreme heat is defined by periods of one day or longer when Maximum Wet Bulb Globe Temperature ($WBGT_{max}$, see page 19 for explanation) > 30 °C. Copied with no changes under Creative Commons Attribution-Non Commercial-No Derivatives 4.0 International (CC BY-NC-ND 4.0) permissions.

residents, over 60% of African urban dwellers (Global Center on Adaptation 2022). Informal settlements usually feature unplanned, densely built structures made with low-cost materials, and are also often established near streams and riverbanks, on hillsides, or in other interstitial spaces (Adegun 2019; Wijesinghe and Thorn 2021). They lack basic services and infrastructure including water and sanitation, and many houses offer little to no heat protection, sometimes acting as 'heat traps' themselves, thereby leaving informal settlement dwellers even more susceptible to the impacts of extreme heat (Jagarnath et al. 2020; Van de Walle et al. 2022).

Informal settlements exemplify the social trends taking place in cities throughout SSA. Rural to urban migration and natural population grown are contributing to rapidly growing and densifying urban areas, often driving out vegetation and green space through LULCC and increasing the UHI. Total exposure to extreme heat is increasing within SSA cities, and particularly within informal settlements, driven not only by social trends but also by environmental context.

Chapter 2
Social and Environmental Context of Extreme Heat

2.1 Climate Trends and Projections for Africa

Global warming has become an irreversible trend. IPCC's sixth assessment report found that from 2010–2020, the global surface temperature had reached 1.1 °C above pre-industrial levels, and has a 50% or greater likelihood of reaching the internationally critical threshold of 1.5 °C of warming in the near future(Mukherji et al. 2023). However, climate impacts such as the rate of warming are unequal across Earth's continents and regions. Africa has experienced rising temperatures at a faster rate than global scales, the rate of increase reaching double the global average in some areas (Li et al. 2022a; Ringard et al. 2016).

Both average and extreme temperatures in Africa are projected to increase as a result of climate change due to both natural and anthropogenic activities (Parkes et al. 2022). Africa's diverse and highly variable climate has already begun to experience altered patterns of precipitation and temperature, including extreme temperature increases recorded across most of the continent (Li et al. 2022a). This has come with many associated costs, including economic (for example, increased energy consumption), agricultural (due to lower agricultural output), and human health tolls. More warming is still to come, though the extent of the temperature increase, and the associated impacts, is uncertain.

Extensive modelling has already been undertaken to predict the realm of possible outcomes. Large differences in potential warming are projected for Africa due not only to the changes in atmospheric composition, but also positive feedback associated with precipitation. Modelling by Serdeczny et al. (2017) found that for subtropical southern Africa, for example, RCP8.5[1] led to a projected decrease in precipitation, thereby limiting the evaporative cooling potential of the soil. The tropics in particular may experience a trend of normalized warming (distinct from historical fluctuations),

[1] Representative Concentration Pathways (RCP) 2.6, 4.5, 6.0, and 8.5 are widely used to predict climate change impacts under low (2.6), medium, and high (8.5, frequently referred to as 'business-as-usual') emissions scenarios.

© The Author(s), under exclusive license to Springer Nature Singapore Pte Ltd. 2025

R. Perrault et al., *Sub-Saharan African Sustainable Development*, City Development: Issues and Best Practices, https://doi.org/10.1007/978-981-95-1458-8_2

with the monthly temperature distribution shifting by more than six standard deviations. Even under a low-emissions scenario (RCP2.6), some regions in western tropical Africa would experience normalized warming up to approximately four standard deviations (Serdeczny et al. 2017).

This aligns with projections reported by Wright et al. (2019: 2), which found that the continent would see "significant increases in the number of days when health may be adversely affected by increasing maximum temperatures due to climate change." This indicates that in under any emissions scenario at least some parts of Africa, if not the entire subtropical region, may need prepare to experience a new climatic regime. This warming, and the associated increase in heat stress, could render some areas uninhabitable by the end of the century due to excessive heat (Parkes et al. 2022).

2.2 Climate Impacts and Drivers

The effects of extreme heat and other climate impacts such as increased precipitation or changes to seasonal weather patterns are substantial and can vary widely. In the physical environment, Africa's Sahel region, for example, has already experienced increasing physical impacts such as tree mortality, stress on food systems, and water shortages, and human impacts including higher rates of heat stress, malaria, and diarrheal diseases. Epule et al. (2017) attribute these impacts to the combination of both rising temperatures as well as rainfall variations in the region. Worku (2018) identifies an even greater range of potential impacts of overheating for the continent and its inhabitants, including increased demand for energy-intensive cooling and strain on energy infrastructure,[2] decreased water quality, damage to temperature-sensitive infrastructure, and increased fire risk for green spaces. Moreover, the human impacts of overheating can be dire.

Often, extreme heat is linked to climate change, but infrastructure, living conditions, and building materials can all act as significant contributors to thermal environments, particularly in cities (Laue et al. 2022). Within cities, other environmental factors include weather conditions; altitude; the size of and geometric configuration of buildings, streets, and the city itself; and density (Feyisa et al. 2014; Worku 2017). Anthropogenic heat contributions such as through fuel combustion, as well as larger-scale behavioural drivers including deforestation, population movements, and land degradation also act as non-climatic drivers of important climate impacts, including extreme heat (Epule et al. 2017).

[2] Refer to Silver (2016) for investigation of energy infrastructure and instability in Accra, Ghana.

2.3 Extreme Heat

Although much of the literature refers to temperature extremes, and heat extremes in particular, extreme heat has an associated set of climatological and functional definitions. The Potsdam Institute for Climate Impact Research and Climate Analytics (2013) uses the historical variability (captured by the standard deviation, sigma) of the current local climate, with the base period of 1951–1980 (Potsdam Institute for Climate Impact Research and Climate 2013). 3-sigma extreme heat events have historically been highly unusual, but over most continental areas are projected to become the norm by the end of the century.[3] Similarly, although many definitions of heat waves can apply depending on the study region, there is no universally accepted definition of a heat wave. Ringard et al. (2016) offer a simple explanation of heat waves as "a period of consecutive days where conditions are excessively hotter than normal at that time of the year" (p.67).

Other organizations use absolute thresholds for understanding more practical applications of the concept and consequences extreme heat. The US National Weather Service issues an 'excessive weather warning' for a period with a Maximum Heat Index, or HI_{max}, in excess of 40.6 °C for two or more days (Tuholske et al. 2021). The International Standards Organization (ISO) uses a Wet Bulb Globe Temperature maximum ($WBGT_{max}$)[4] of greater than 30 °C to define the occupational heat stress threshold for risk of heat-related illness. This threshold has been associated with higher mortality rates among vulnerable populations.

Extreme heat and heat waves are increasing globally, and in Africa in particular. Globally, urban exposure to $WBGT_{max}$ of > 30 °C increased almost 200%, measured in person-days per year over the last several decades (Tuholske et al. 2021). The increasing number of days that meet the occupational heat stress threshold and/or excessive weather warning threshold may occur at a faster rate in Africa than elsewhere in the world (Wright et al. 2019). Nonetheless, the IPCC has predicted with high confidence that days with heat extremes in Africa will increase under both low- and high-emissions scenario (Portner et al. 2022).

The increased frequency of extreme heat and heat waves driven by both natural and anthropogenic factors and rising temperatures will have substantial impacts on urban environments and could inhibit livability for residents. Global and urban warming has increased immensely in recent decades, and although they do not always leave physical evidence, this climate impact will be a consequential challenge to SSA cities.

---

[3] Meanwhile, 5-sigma events, which under a normal distribution would have a return time of several million years, have "almost certainly never occurred over the lifetime of key ecosystems and human infrastructure" (Potsdam Institute for Climate Impact Research and Climate 2013: 12).

[4] $WBGT_{max}$ is a measure of heat stress in direct sunlight that captures the biophysical response of temperature-humidity combinations (US Department of Commerce n.d.).

Chapter 3
Physical Impacts of Extreme Heat on Urban Environments

3.1 Challenges Faced by African Cities

Aside from increasing urban heat, cities in SSA face numerous interconnected challenges that are dependent on a variety of factors. Among others, many cities share three key challenges, namely, (1) poor (or lack of) planning; (2) rapidly growing populations (and the institutional challenges associated with accommodating them); and (3) a large proportion of informal settlement dwellers and/or urban residents living in poverty,[1] among others (Hunter et al. 2020; Lindley et al. 2018). These pressures combine to create a situation where economic growth and infrastructure cannot keep up with urban expansion and development leaving many residents underserved and vulnerable to climate hazards and other risks (Cavan et al. 2014; Global Center on Adaptation 2022). Poor planning, growing populations, and informality and poverty makes the provision of basic infrastructure, let alone effective climate adaptation, very challenging.

The predominant planning practices and other measures to reduce urban insecurities that are used in urban SSA have often failed or been insufficient for addressing the issues at hand (Titz and Chiotha 2019; Worku 2017). Criticisms of failed planning regimes sometimes point to their tendency to exist as imported measures: either from other continents or contexts, or as legacies of colonial regimes (Lindley et al. 2018).

The implementation of new urban green infrastructure (UGI) faces numerous barriers to implementation, including financial constraints (Thorn et al. 2021). Nonetheless, the potential of planning measures to reduce the UHI effect has been well-established. Li et al. (2022a) point to morphological factors that can be established through top-down planning such as wind corridors, which have long been used in the construction styles of many coastal and arid zones to leverage passive cooling

---

[1] In part driven by economic growth: when the prosperity of the urban African middle and upper classes is on the rise, this growth drives demand for energy, transport, and housing infrastructure, in turn contributing to the growth of informal and working-class neighbourhoods to provide services and respond to new demands.

© The Author(s), under exclusive license to Springer Nature Singapore Pte Ltd. 2025
R. Perrault et al., *Sub-Saharan African Sustainable Development*, City Development: Issues and Best Practices, https://doi.org/10.1007/978-981-95-1458-8_3

for local sun and wind conditions, and applications of UGI to regulate urban temperatures. But this can lead to the assumption that formality through planning would lead to higher service levels, thereby justifying policies that aim to convert high-density informal settlements to other housing types (Cavan et al. 2014). However, for ecosystem services, this is not necessarily the case. The provision of vegetation and green space, and associated temperature reductions, is highly context-dependent.

Thus far, governments have mostly not been successful in creating sustainable master plans for cities that could contribute to building resilience for all residents in the face of climate hazards, and moreover have often failed to provide the basic amenities of life to residents, particularly to residents of informal settlements (Salami et al. 2018). Local governments lack "expertise, funds, and data on urbanisation [which] continues to hinder efforts to improve living conditions and implement sustainable natural resource use" (Wijesinghe and Thorn 2021: 1).

The growing urban populations of sub-Saharan African cities has led to substantial institutional challenges in accommodating and adequately supporting the growing urban populations and city boundaries. Government and governance play a substantial role in both creating and addressing what has been described as a "wide range of interwoven institutional challenges – and thus an associated 'multidimensional accumulation of risk'" (Titz and Chiotha 2019: 7).

Weak political institutions have been identified as a major contributor to urban vulnerabilities, alongside poor governance. Governance is not limited to state actors and formal institutions, but existing institutions have very limited capacity to enact planning strategies and bylaws. Titz and Chiotha (2019: 6) identify one potential reason for the institutional weakness of governments being the "hasty and only partial decentralisation of public administration in Africa since the 1980s, which has led to local governments characterized by weakness, disorganisation, inadequate training and staffing, and overwhelming burdens from new, complex tasks". With serious concerns such as providing basic services to growing populations, including in informal settlements, there is a well-documented lack of institutional capacity to prioritize less-visible concerns such as the UHI and less pressing concerns like lack of green space (Lindley et al. 2018).

While state governance is limited, informal practices govern the lived realities of the city. Informal settlements are where the most unregulated growth is taking place, and the growth patterns and socio-spatial relationships naturally emerge from the bottom up (Jagarnath et al. 2020; Lindley et al. 2018; Titz and Chiotha 2019). These communities are sensitive to climate impacts including increased thermal discomfort and heat stress due to low adaptive capacity and the conditions of the local built environment (Hunter et al. 2020; Laue et al. 2022). Housing infrastructure in informal settlements is often made of cheap and lightweight materials that conducts and traps heat, thereby not offering any protection to heat and in some cases, making shelters hotter than the outdoors (Muhammad and Wonorahardjo 2020; Van de Walle et al. 2022). The increased risk of heat exposure in the informal settlement urban morphological type has been confirmed, and as such, extreme heat has significant human health impacts in informal settlements (Adegun et al. 2022). However, as mentioned above, service delivery and infrastructure backlogs are seen as more pressing concerns for these areas, making heat risk planning a lower priority for already-stretched municipalities (Jagarnath et al. 2020). In addressing these high-priority actions, integrated design strategies such as planning for infrastructure and UGI alongside housing and transport would be of great benefit.

3.2 Physical Impacts of Extreme Heat

3.2.1 Extreme Heat in Urban Environments

In the context of warmer temperatures, an increased frequency of extreme events, and rapid urbanization, the urban thermal environments of SSA's cities are shifting dramatically. Factors like congestion, urban expansion, industrialization, and inadequate management can all contribute to a higher risk of climate disaster for urban areas, which, along with increased exposure to extreme heat, can threaten the sustainability of urban settlements. One of the most important factors is the UHI effect, which is an anthropogenic climate modification that has been observed in some capacity in cities for centuries (Oke 1982; Yow 2007).[2]

Amid growing concern about the impact of humans on the environment, there has been a robust scholarship on the UHI's causes, impacts, and adaptations since the 1950s. Urban temperatures and thermal environments are clearly different from rural ones, and heating and cooling patterns modified by urban environments are theorized to be responsible (Oke 1982; Yow 2007).

The UHI can raise the temperatures within cities by several degrees, depending on factors such as pre-existing geography, density, and scale and type of alteration of the natural landscape (Orimoloye et al. 2019). The UHI results from these factors, in combination with other physical characteristics of a city, including the thermal and physical properties of construction materials, building morphology, and anthropogenic heat sources such as vehicle exhaust and industrial activities (Yow 2007).

Urban morphology[3] shapes the UHI and interacts with externally-driven heat events at the city and neighbourhood scale (Worku 2017). Certain types of development or densification can create wind tunnels in some cases, or reduce or cut off fresh air flows in others, as is more often the case for uncontrolled development patterns (Lindley et al. 2018). For example, consider a wide, tree-lined boulevard with well-built white houses and large gardens in comparison to a network of small alleys navigating densely built shacks made of corrugated metal, with bare ground and little green space in sight. Each urban morphology type, with its unique spatial

[2] As discussed in the introduction, the UHI is defined as "the phenomenon of higher urban temperatures than the rural surroundings and affects temperatures in the air and at [land surfaces]" (Li et al. 2022a: 4).

[3] Morphology in this case is used in the architectural sense of the term, meaning the physical forms such as shape, layout, mass, and surface cover of the built environment.

and thermal characteristics, can help explain the gravity of climate impacts such as extreme heat on its residents, and along with the socioeconomic status of the district, impacts on residents and their adaptations (Cavan et al. 2014; Van de Walle et al. 2022).

LULCC during the urbanization process including increased densification and urban sprawl can also increase the intensity and magnitude of UHIs, and exacerbate heat stress experienced by urban dwellers (Li et al. 2022a). Rapid urbanization can lead to deforestation and reduction in vegetation cover, in favour of concrete, asphalt, and other infrastructure and building materials. These materials (i.e. dark and impervious surfaces) can worsen the UHI due to their thermal properties of low specific heat and high heat capacity (Muhammad and Wonorahardjo 2020). Moreover, surface sealing can contribute to or worsen other disasters, for example, by increasing surface runoff of rainwater (Titz and Chiotha 2019).

LULCC has significant social, economic, and environmental impacts on urban dwellers and ecosystems (Di Leo et al. 2016). The combination of land cover types present in an urban area or city as a whole can impact temperatures and the UHI (Cavan et al. 2014). Limited green space within cities has been linked to stronger UHI impacts (Worku 2017). In many sub-Saharan African cities, access to green space and ecological services from it is under threat and declining rapidly (Adegun et al. 2022; du Toit et al. 2018). For example, as Abuja, Nigeria underwent rapid urbanization between 2001 to 2006, its land area dedicated to green space declined from 21% to a mere 12%; in Kumasi, Ghana, the expansion of the urban boundaries to expand the city size from 182km^2 in 1963 to 254km^2 in 2011 resulted in massive destruction to the city's green belt and other core green spaces (Mensah 2014). These declines contribute to the UHI which, along with climate change, is leading to higher temperatures in cities.

3.2.2 *The Human Impacts of Extreme Heat*

The climate and climate events can have profound impacts on human health, ranging from acute and immediate symptoms to chronic illness, and even death. As near term changes in climate parameters including humidity, temperature, radiation, and wind take place, the impacts of climate events on human health could experience sharp rises (Orimoloye et al. 2019).

In Africa and globally, studies have found higher temperatures to be associated with increased mortality (Tuholske et al. 2021; Wright et al. 2019). Tuholske et al. (2021: 3) reported that "globally, for every additional day that [the maximum air temperature] exceeds 35 °C compared to 20 °C, mortality increases by 0.45 per 100,000 people, with an increase of 4.7 extra deaths per 100,000 people for those above 64 [years old]." Another study investigating all-cause mortality in northern Ghana found that there was a statistically significant association between mortality and mean daily temperatures above the 75th percentile, highlighting the importance of short-term weather variability and air temperatures in particular on human health (Azongo et al. 2012). Mortality due to extreme heat varies by location and climate zone, however the health risks of heat are more acute in densely populated areas (Orimoloye et al. 2019; Serdeczny et al. 2017).

Heat-related morbidity is also of concern. Higher temperatures have been associated with morbidity from cardiovascular, respiratory, and cerebrovascular diseases, diabetes Mellitus, and renal disease (Wright et al. 2019). Other heat-related illnesses include diarrhea, sunburn, heat stroke, and skin cancer, while human bodily responses to heat, may also include dehydration, sweating, heat rash, heat exhaustion, headaches, and sleep disturbances. Increasing incidences of these symptoms due to extreme heat have also led to increasing hospital admissions in the continent, collectively linked by the same likely cause: heat stress (Li et al. 2022a; Ncongwane et al. 2021; Orimoloye et al. 2019).

Heat stress arises from exposure to extreme heat and occurs when a person's body is unable to maintain homeostasis, leading to abnormally high internal body temperatures (hyperthermia). In other words, heat stress is experienced when the body is overheating. Heat stress can lead to the symptoms described above, while less severe effects can include sweating, headache and nausea, weakness, fatigue, and dizziness (Wright et al. 2019). The impacts of heat stress are not limited to physical factors, as negative dietary effects (such as reduced caloric intake and underweight status) and psychological symptoms and productivity decreases are common too (Mueller and Gray 2018).

Numerous other health impacts are possible from extreme heat. Extreme heat and heat wave events can increase the degree of atmospheric pollution and increase the production of certain air pollutants and allergens (Ncongwane et al. 2021; Ringard et al. 2016). Numerous respiratory and cardiovascular diseases can be caused by air pollution. Extreme heat and increased temperatures may also have impacts on human skin, the body's largest organ. Exposure to excessively high temperatures may cause damage to the skin, and increased ultraviolet radiation introduces risks such as heat stroke, skin cancer, and heart disease. Additionally, increased incidence of non-cholera diarrhea may be correlated to increased ambient temperatures, indicating that heat is a weather variable that may contribute to the increased occurrence of gastrointestinal illness, which is a major cause of morbidity and mortality in developing nations (Li et al. 2022a; Orimoloye et al. 2019).

There are several risk factors that influence the likelihood of an individual experiencing these impacts. Children, the sick, and the elderly are in broad terms the most vulnerable to the adverse effects of extreme heat (Li et al. 2022a). Wright et al. (2019) found in their South Africa-based study that age is a statistically significant factor in a person's likelihood to experience heat-health effects. Their findings were aligned with their other studies reviewed, which showed that "people above 60 years are at increased risk of heat-related illness because of physiological impairments in the regulation of core body temperature in hot conditions", resulting in people older than 60 being almost twice as likely to experience heat-health effects than younger individuals (Wright et al. 2019: 4). Moreover, in line with findings from the Euro-HEAT project, Wright et al.'s data showed women to be 1.15 times more likely to experience heat health effects than men, possibly due to comparatively more time spent indoors to care for children and the elderly and to conduct housework.

Socioeconomic factors are also highly relevant. People living in urban environments, in poverty, or both are most vulnerable to heat risks in Africa, but exposure to extreme heat within urban settlements is unequal too and disproportionately impacts the urban poor (Tuholske et al. 2021). There is an association between unplanned urban development and UHI: for instance, higher air temperatures have been observed in informal settlements compared to other areas in SSA cities, and inhabitants of these areas, as well as poor suburb areas, are generally more vulnerable to hazards due to high-risk living conditions and limited access to basic services (Li et al. 2022a; Orimoloye et al. 2019). Ncongwane et al. (2019: 2) offer an apt summary of the issue: "ongoing socio-economic challenges have been found to worsen the interdisciplinary facets of extreme heat vulnerabilities stemming from the changing climate."

Increased heat exposure is not going unnoticed to the groups most vulnerable to it. In a study by Adegun et al. (2022), 79% of 405 residents of an informal settlement surveyed in Dar es Salaam, Tanzania, agreed that their neighbourhood had become hotter in recent years. Causes identified by residents included densification, house site and form, sanitation practices, and absence of green structures. Taiwo's (2012) study of low-, medium- and high-density neighbourhoods in Ibadan, Nigeria, also found that one of the most important factors affecting heat exposure was lack of neighbourhood trees, along with number of electricity hours and distance from water supply points, both of which can be precarious in informal settlements. Another study pointed to lack of tree shade as a significant cause of heat exposure in Ghanaian informal settlements, alongside housing material (Wilby et al. 2021).

The linkages made between heat exposure and trees or vegetation are not unfounded. In a study with over 40 temperature-measuring points located within different morphology types across Kampala, Uganda, Van de Walle et al. (2022: 1) found that "Normalized Difference Vegetation Index (NDVI)[4] has the highest power (75%) to predict the intra-urban variations in daily mean heat stress", variations that found a daily maximum difference of 6.4 °C between different stations in the observation period, reaching 14.5 °C of difference on the most extreme day.

Thus, it is clear that while heat stress is usually presented as a health challenge in existing literature that focuses on the African context, the challenge of heat exposure reaches far beyond that (Laue et al. 2022). There are many factors that determine heat exposure, including but not limited to "hours of electricity supply, distance from water supply points, number of trees around the house, average monthly income,

[4] A measure used to estimate land surface emissivity and to differentiate vegetation from other land cover types.

number of people with a minimum of primary school education, housing ventilation status, building density, aggregate mean distance separating buildings, daily maximum hours spent under the sun and private car ownership status" (Taiwo et al. 2012: 410). As such, heat exposure is not merely an issue of climate and geography, nor of individual risk factors but is in fact deeply tied to socioeconomic status and historical spatial patterns of inequity.

Chapter 4
Responses to Extreme Heat

Climate impacts necessitate responses. Adaptive responses are actions or modifications to behaviour or environments undertaken to adapt to circumstances, such as excessive heat and other climate impacts. Responses are often based on past successes and failures, frequently leveraging local knowledge, available resources, and creativity, depending on the stimuli. Over time, the number of adaptation actions reported and studied has been increasing, at both continental and regional scales (Epule et al. 2017; Hunter et al. 2020). However, whether reported on or not, responses to climate stimuli are almost certain to take place.

Sometimes, responses are fairly unconscious (for example, opening a window to allow air circulation when it is hot out), and sometimes they occur at a more systematic level (for example the implementation of a city climate adaptation action plan). While the number of reported adaptations is not indicative of success, impact, or resilience, the type and frequency of such actions are important indicators of the adaptive capacity of a society. This includes whether responses take place in response to short-term motivations or long-term trends, and the actors undertaking these responses.

Types of adaptive responses include situational modifications such as income diversification for livelihoods under threat, or behavioural modifications such as taking a shower and wearing light clothing in response to heat exposure (Adegun et al. 2022; Epule et al. 2017). Adaptive responses can be for immediate or short-term relief, as well as in response to longer-term trends, such as the construction or reconstruction of physical structures or to improve drainage (Hunter et al. 2020). Adaptive responses can be taken by different actors, including individuals or households, and at the municipal and state level, and can be anticipatory, reactive, or both.

© The Author(s), under exclusive license to Springer Nature Singapore Pte Ltd. 2025
R. Perrault et al., *Sub-Saharan African Sustainable Development*, City Development: Issues and Best Practices, https://doi.org/10.1007/978-981-95-1458-8_4

4.1 Reactive Responses

Reactive responses take place during or after a climate event and are done in reaction to its impact. Most documented responses to climate drivers are reactive, and many can be thought of as "coping" (Hunter et al. 2020). Although there are many potential responses to extreme heat or higher temperatures, in a systematic review of responses to climate impacts in urban Africa, Hunter et al. (2020) found that responses to extreme temperatures were mostly (81%) reactive, with the goal of adaptation to being adapting to or coping with the existing circumstances.

Adaptation to extreme heat is not as well-researched as adaptation to other climate impacts and drivers, with substantial focus in the existing literature oriented towards acute hazards and shocks such as flooding (Laue et al. 2022). However, heat adaptation can include a wide variety of measures and techniques including behavioural responses as well as economic adjustments which involve spending money to improve comfort. Behavioural responses are reactive responses that include changes to behaviour due to the presence of a certain impact. Behavioural responses to heat stress and thermal discomfort in informal settlements include bathing/showering and wearing light clothing, as previously mentioned, as well as reducing activity levels, opening windows and doors to improve indoor air flow, drinking water or other chilled fluids, using hand or electric fans, and wearing loose, light clothing (Adegun et al. 2022; Laue et al. 2022).

Responses involving economic adjustments could include purchasing and using air conditioners, purchasing new, lighter clothing, or staying out late in cooled restaurants and hotels (Taiwo et al. 2012). These economic options are mostly available to more affluent groups who can afford air conditioners (Parkes et al. 2022). This adaptation strategy is economically out of reach and therefore uncommon for informal settlements and low-income housing.

Adapting to heat can involve modifications to the built environment too, often leveraging opportunities to reduce the UHI. Shielding solar radiation through awnings, morphological architectural features for passive cooling, reducing energy consumption and incorporating vegetation in developed areas have been found to be effective ways of improving thermal environments (Takebayashi et al. 2020; Yow 2007). Other response strategies can include establishing fresh air corridors that are as long and wide as possible; reducing ground-level temperatures through effective building orientation, increased vegetation, inclusion of open water surfaces, and encouraging ventilation through diverse building heights; and integration of architectural measures that reduce heat load and manage solar radiation including white reflective or green vegetated roofs and facades and shading (Worku 2017). While it is difficult to retroactively modify well-established urban forms, installing trees and vegetated surfaces is also frequently mentioned as a potential way to reduce the UHI within certain existing morphological constraints.

4.2 Anticipatory Responses

Anticipatory responses are not generally as common as reactive responses, as they involve planning in advance and are not usually undertaken without sufficient adaptive capacity. This is unfortunate, as adaptive measures undertaken in advance can often have much more significant impacts and returns on investment (financial and otherwise) than reactive responses (Global Center on Adaptation 2022). In Hunter et al.'s (2020) systemic review, the authors found only 8% of studied adaptation actions were taken explicitly in anticipation of long-term climate change. Almost half of such responses implemented by local governments, a positive sign because effective planning (especially at the municipal level) can maximize the impacts of anticipatory responses, in particular, by targeting specific efforts towards addressing community vulnerabilities (Jagarnath et al. 2020).

UGI could be an impactful anticipatory response, including in informal settlements where, as Laue et al. (2022: 8) explain, "the prevalence of heat and thermal discomfort is linked to a low green space coverage [and] the presence of vegetation is a significant predictor of cooler mean temperatures and lower perceived temperatures within informal settlements". UGI takes some time to implement, and the relief it offers not immediate, but its payoffs can be significant and are recognized by many communities. In fact, implementing UGI to address heat stress is already well-documented in some informal settlements, ranging from growing creepers on the sides of buildings to regulate temperatures, planting of trees by communities, and implementing vertical greening systems (Laue et al. 2022).

Chapter 5
Urban Green Infrastructure Solutions

Urban green infrastructure (UGI) and nature-based solutions (NBS) can play a pivotal role in urban heat regulation. NBS are defined by the European Commission as follows:

> "[NBS are] solutions that are inspired and supported by nature, which are cost-effective, simultaneously provide environmental, social and economic benefits and help build resilience" (Voskamp et al. 2021: 2).

Implementing and properly maintaining UGI is one such solution. Although different authors offer variations on definitions, for the purpose of this research, UGI will be understood as follows:

> "[UGI are] all natural, semi-natural and artificial networks of multifunctional ecological systems within, around and between urban areas, at all spatial scales" (Tzoulas et al. 2007, as quoted by du Toit et al. 2018: 1).

Wijesinghe and Thorn's (2021: 2) definition of UGI as "'permeable, multi-functional and interconnected spaces that support environmental, recreational and wellbeing functions within a city, through provisioning, regulating, supporting and cultural ecosystem services" is also useful, linking the concept of UGI to ecosystem services. UGI includes urban areas covered by vegetation, and for the purpose of this research can scale down to include discrete green elements such as trees, or scale upwards to be understood as broader systems or networks that work with engineered, 'grey' infrastructure (Thorn et al. 2021; Wijesinghe and Thorn 2021). One type of UGI is urban green spaces (herein referred to as simply "green spaces"), which are defined as:

> "An umbrella term for all areas of land [that consist predominantly of unsealed, permeable, 'soft' surfaces such as soil, grass, shrubs and trees]…whether or not they are publicly accessible or publicly managed. It includes… all areas of parks, play areas and other green spaces specifically intended for recreational use, as well as other green spaces with other origins" (Swanwick et al. 2003, as quoted by du Toit et al. 2018: 1).

© The Author(s), under exclusive license to Springer Nature Singapore Pte Ltd. 2025

R. Perrault et al., *Sub-Saharan African Sustainable Development*, City Development: Issues and Best Practices, https://doi.org/10.1007/978-981-95-1458-8_5

These concepts have grown substantially in popularity and recognition in recent years (Lindley et al. 2018; Titz and Chiotha 2019). NBS in the form of UGI and green spaces has been recognized as a key opportunity area for the development of climate-resilient urban areas in recognition of NBS and UGI's substantial potential to help cities adapt to climate change (African Union Commission 2022).

5.1 Ecosystem Services

The concept of ecosystem services is also growing in importance, particularly with respect to UGI. Ecosystem services offer several roles and benefits, including provisioning (e.g. fuelwood, food), regulating (e.g. temperature, flood), supporting (e.g. habitat, water filtration), and cultural (e.g. recreational, aesthetic) services, thereby increasing the resilience of cities to climate change (Cavan et al. 2014; Wijesinghe and Thorn 2021).

UGI offers improved ecosystem services such as flood risk reduction, reduced UHI and air purification to cities, and serving to support biodiversity, all while potentially contributing to bio-physical and social wellbeing for residents (Nassary et al. 2022; Titz and Chiotha 2019). UGI can also support food security and income generation, and substantial research exists on how it can benefit all parts of society.

These services are of great importance to all societies but play a particularly important role for groups with low adaptive capacity and high vulnerability, such as residents of informal settlements. One study of low-income dwellers in hazard-prone communities of Dar es Salaam, Tanzania, found strong dependencies on regulating and provisioning ecosystem services for both livelihoods and resilience (du Toit et al. 2018). Other studies "clearly established the fact that low-income households directly depend on the natural resource environment for their lives and livelihood" (Adegun 2019: 78).

UGI has been shown to have a desirable impact on urban temperatures, and is frequently cited as a potential mechanism for reducing the UHI (Di Leo et al. 2016; Drescher 2019; Li et al. 2022b; Van de Walle et al. 2022; Wolch et al. 2014). Vegetation enhances transpiration, which leads to air temperature cooling. The efficacy of UGI's temperature regulation is determined by several factors, including the physical characteristic of the UGI, the quality of the green structure, and the type and density of vegetation (Cavan et al. 2014; Feyisa et al. 2014; Taiwo et al. 2012). Indeed, trees and vegetation can contribute to substantial variability in heat impacts. For example, Cavan et al. (2014) found in their comparative study of Dar es Salaam, Tanzania and Addis Ababa, Ethiopia that despite Dar es Salaam's higher air temperatures, its modelled surface temperatures are lower due to that city's higher evaporating fraction.[1] Temperature regulation is a critical ecosystem service that is relatively well-documented, with du Toit et al. (2018) finding it to be one of the most assessed

[1] Fraction of vegetated and water surfaces.

ecosystem services in their systemic review of peer-reviewed literature on urban green infrastructure and associated ecosystem services in sub-Saharan African cities.

5.2 Urban Green Infrastructure in Informal Settlements

UGI can have a significant impact on mitigating the UHI and positively impacting residents of informal settlements through temperature regulation, pollution reduction, carbon storage, wildlife or biodiversity habitat, and improved human health and well-being (du Toit et al. 2018; Feyisa et al. 2014; Zhang et al. 2021). Green space is an important part of the sustainability and liveability of cities worldwide, and was historically implemented in sub-Saharan African cities in the colonial and post-colonial era, but these spaces are gradually losing out to processes of urbanization and densification in this period of rapid population growth and associated pressures (Titz and Chiotha 2019; Wijesinghe and Thorn 2021).

Due to their generally unplanned nature, UGI is not always formally incorporated into informal settlements, but plays an important role in these areas nonetheless. In the informal settlement context, Adegun (2019) observed that having a domestic garden was an aspirational goal for many residents of Kya Sands settlement in Johannesburg, South Africa. The gardens within the settlements (ranging from potted plants in outdoor containers to plots of a few square meters) shown in Fig. 5.1 below, contributed to household food consumption, aesthetics, and micro-climate control benefits. Similarly, Wijesinghe and Thorn (2021) found that UGI is valuable to informal settlement residents in Windhoek, Namibia, with riparian vegetation adjacent to the settlement, meeting areas, sports grounds, hilly slopes, urban farms, garden plots, and green fences offering provisioning, regulating, supporting, and cultural ecosystem services.

However, provision of green space has repeatedly been linked to patterns of exclusion and gentrification in a variety of global contexts (Titz and Chiotha 2019; Wolch et al. 2014). Even in informal settlements today, the historical link between green spaces and patterns of exclusion is replicated, for example charging users to enter resident-developed 'park' spaces and forbidding use by women (Adegun et al. 2022). This pattern results in the inequitable distribution of the benefits of UGI, including ecosystem services, restricting access from those who need them most.

Fig. 5.1 Some of the domestic gardens in Kya Sands Settlement. Photographs by O.B. Adegun, May/June 2014 (Adegun 2019)

Chapter 6
Dimensions of Vulnerability and Potential Applications of Urban Green Infrastructure

Africa is one of the most vulnerable continents to climate change. This is due to two main factors: adaptive capacity and geographic characteristics. Africa's geography is described by Busby et al. (2013: 136–137) to be such that "the physical effects of climate change are likely to be the most severe on the planet". Within the continent, regional differences in physical and climate vulnerability exist. East Africa is often seen as one of the most climate-vulnerable regions at present, while tropical West Africa is projected to experience the most significant temperature changes (Li et al. 2022a; Serdeczny et al. 2017). However, these are not to be viewed in isolation from other dimensions of vulnerability.

Less developed countries are generally considered to be more vulnerable than developed countries, and poor individuals more vulnerable than the rich to suffering negative impacts from hazards and risks (Taiwo et al. 2012). There is no exception when it comes to extreme heat. Spatial mapping of social vulnerability and heat stress risk in Durban by Jagarnath et al. (2020) shows that both social vulnerability and heat vulnerability exhibit similar spatial patterns, and the authors suggest that the latter is actually a function of the former. The intersection of urban expansion, climate change, and inhabitation of certain housing types such as informal settlements inequitably distributes the heat-related disease burden and mortality to marginalized and low-income groups (Adegun et al. 2022). The inequitable distribution of vulnerability is apparent in many facets of the challenge of urban heat.

6.1 Energy and Air Conditioning

In the context of extreme heat in particular, energy infrastructure and instability in Africa could represent a major source of vulnerability for some cities. In an analysis of the UPE of interrupted electricity in Accra, Ghana, Silver (2015) discusses the frequent disruptions to the city's electricity network in terms of power relations of

© The Author(s), under exclusive license to Springer Nature Singapore Pte Ltd. 2025

R. Perrault et al., *Sub-Saharan African Sustainable Development*, City Development: Issues and Best Practices, https://doi.org/10.1007/978-981-95-1458-8_6

Accra's residents and the city's socio-natural relations as a whole. Wealthy urban groups are well-insulated from the impacts of heat, with access to air conditioning, mechanical fans, and backup generators for their homes that keep such cooling devices working regardless of the reliability of electricity services.

Meanwhile, energy demand continues to rise throughout the city, with infrastructure not keeping up, particularly in informal settlements and other neighbourhoods housing the urban poor, who rely on energy for their livelihoods and have low adaptive capacity and therefore a limited selection of potential response strategies (Silver 2015). Moreover, air conditioners can be considered a maladaptive technological response: aside from driving blackouts or other energy system disruptions, they are known to increase urban heat, intensifying the UHI and energy demand, and posing a threat to energy stability in some urban centres and regions while exacerbating the issue for those without reliable access to energy and worsening the impacts of the UHI effect on poorer residents (Marks and Connell 2023).

Although residential and other energy demand is necessary to maintain services and indoor thermal comfort needed to limit health issues from heat events, during heat waves demand could reach unsustainable peaks, outstripping supply at the very same important moment. Under all combinations modelled by Spinoni et al. (2017), increased population-weighted cooling degree-days (a measure of the estimated energy required to cool the environment) and population itself will lead to increased energy demand in the coming decades. Globally, the increase in population-weighted energy (cooling and heating) degree-days was modelled to be the largest in tropical latitudes, particularly in western and central-eastern Africa (Spinoni et al. 2021). The increased global energy demand could lead to increased stress on energy infrastructure and supply disruptions, and increase vulnerability of populations to the health and other impacts of extreme heat events (Li et al. 2022a; Parkes et al. 2022).

UGI could have the potential to reduce vulnerability to both household heat discomfort and energy demand for cooling overall. In areas with UGI, surface temperatures have been found to be consistently cooler than adjacent areas without UGI and can create positive temperature anomalies, especially in areas of high canopy cover (Di Leo et al. 2016; Drescher 2019). By reducing temperatures in wealthy areas, electricity demand for air conditioning on hot days could decrease, thereby assisting to increase the energy network's resilience. UGI in low-income neighbourhoods including informal settlements could offer thermal regulation ecosystem services, which could help reduce overall vulnerability to heat events, including by reducing the risk of heat stress.

6.2 Agriculture and Food Security

Agriculture is another dimension of vulnerability that is heavily affected by climate impacts and could contribute to other vulnerabilities such as food security and economic wellbeing. The sector employs 65% of Africa's labour force, but production is closely tied to the climate, as rainfed agriculture comprised 96% of crop production as of 2017 (Serdeczny et al. 2017). This means that changes to precipitation patterns and heat waves could pose a major threat to the sector.

Heat waves can have significant impacts on both agriculture and the hydrological processes that impact it (Ringard et al. 2016). Many crops (maize, wheat, and sorghum, for example) have particular temperature sensitivity thresholds above which, large yield reductions are expected (Serdeczny et al. 2017). Lower precipitation, higher temperatures, and more frequent ecological disruption events such as the desert locust outbreak in East Africa in 2019–2020 add to the challenge of direct climate impacts by posing a major threat to food security and household livelihoods (Li et al. 2022a).

Agriculture is another dimension of vulnerability that could potentially benefit from UGI. Research on the advantages and feasibility of UGI for urban agriculture in SSA has already been undertaken, as food is a key provisioning service that UGI could potentially offer (Di Leo et al. 2016; Laue et al. 2022). Considerable differences in land surface temperature between agricultural local climate zones and other zones were found by Li et al.'s (2022a) investigation, possibly due to the proportion of vegetation. Moreover, in informal settlements, UGI in the form of domestic and communal gardens has been shown to have the triple benefit of contributing to thermal regulation, providing food, and strengthening community ties (Adegun 2019).

SSA cities are subject to numerous vulnerabilities, but as demonstrated by these examples, their potential impacts are most acute for the socioeconomically marginalized. In the case of energy instability, air conditioning and other mechanical cooling is inaccessible to those most vulnerable to heat stress but contributes to furthering existing vulnerability. In the case of agriculture, climate change threatens food security, particularly in cities with rapidly growing populations. UGI is posited to be a potential solution in both cases, contributing thermal regulation and provisioning ecosystem services.

Chapter 7
Potential Applications and Challenges

7.1 Vulnerability Dimensions

7.1.1 Distribution of Heat Vulnerability: Informal Settlements

Rapid urbanization and densification and the socioeconomic constraints of the urban poor have subjected these groups, including residents of informal settlements, to higher levels of heat risk. Increasing urbanization and growing populations meet weak institutional capacity and informal governance mechanisms to create socio-spatial relationships shaped by socioeconomic status. Through the UHI extreme heat events are worsened, if not created, by the interactions between social and natural factors; the burden of which is subject to an overlapping spatial distribution with that of poverty.

Many of the most vulnerable groups to heat exposure have the lowest adaptive capacity and inhabit informal settlements. These settlements, densely built and made of cheap and heat-conducive material, have in turn been shown expose residents to higher levels of heat exposure (Adegun et al. 2022; Cavan et al. 2014; Hunter et al. 2020; Jagarnath et al. 2020; Laue et al. 2022; Van de Walle et al. 2022; Wijesinghe and Thorn 2021). At the household level, socioeconomic factors are closely tied to heat vulnerability, heat stress, and heat exposure. Housing and environmental exposure reflect these socioeconomic conditions and heavily influence health risks from extreme heat.

The more powerful socioeconomic groups shape cities, and the balance of power between the urban elites and urban poor, along with scarcity of public resources, influences access to infrastructure, stability of services, and resilience (Silver 2015). While urban elites enjoy private green spaces (possibly including trees for shade and

© The Author(s), under exclusive license to Springer Nature Singapore Pte Ltd. 2025

R. Perrault et al., *Sub-Saharan African Sustainable Development*, City Development: Issues and Best Practices, https://doi.org/10.1007/978-981-95-1458-8_7

air-conditioned homes), densely inhabited informal settlements have little heat-regulating vegetation cover; are often made of heat-trapping material; have UHI-inducing morphological characteristics; and have unreliable access to energy, cooling devices, and other infrastructure services (such as water) for effective cooling. Access to infrastructure includes to UGI, which most of the literature reviewed agrees is better maintained and more accessible in affluent areas and to wealthy social groups, perpetuating historic inequalities. UGI and formal green spaces that provide temperature regulation and other ecosystem services (Adegun et al. 2022; Laue et al. 2022; Thorn et al. 2021; Titz and Chiotha 2019).

7.1.2 Distribution of Spaces of Protection: Urban Green Infrastructure

Some scholars view cities as socio-ecological systems, thereby driving the view of UGI away from being a planned element in the city towards a view of UGI being an element *of* the city, contributing to its overall functioning as a holistic ecosystem (Lindley et al. 2018; Wijesinghe and Thorn 2021). Under this view, the city itself is not outside of nature, but part of it, with urbanization conceived as an ecological process and the socio-natural relationship expressed through it and through the transformation of nature.

"Nature" in the form of green space or UGI is conversely a form of infrastructure, to which, like many other forms of infrastructure in many SSA cities, access is unequal. As the political economy shapes urban fabric, so too does it shape the distribution of UGI and its core functions, be it leisure and added thermal regulatory benefits, or a sparse but essential support to livelihood. As shown in the cases of energy and air conditioning, and agriculture and food security, UGI has the potential to reduce vulnerability, increase equity, and improve heat stress outcomes across urban communities. Access to UGI can reduce vulnerability to heat through its thermal regulation services, for example, and improve community resilience in many other capacities.

However, due to pressures such as density, access to water, and access to financial resources (for maintenance, for example), UGI and its benefits are unequally distributed and often inaccessible to lower-income groups.[1] Many high-level benefits

[1] There is some disagreement in the literature on the relationship between UGI and affluence in sub-Saharan African cities. Most scholars agree that it is similar to Western contexts, wherein high socioeconomic status increases access to green space and therefore better provision of temperature regulation services. However, in their study of urban morphological determinants of temperature regulating ecosystem services in Addis Ababa and Dar es Salaam, Cavan et al. (2014) found little evidence that higher class residential areas did in fact have more access to green space than lower class areas. This stands in contrast to the findings of several other authors, who found evidence of little or poor UGI in less affluent areas, as well as barriers to its maintenance (Laue et al. 2022; Thorn et al. 2021; Titz and Chiotha 2019). However, Cavan et al.'s work only addresses two large cities, limiting potential generalizations that can be drawn from their findings, particularly since each city

of UGI such as thermal regulation and aesthetics continue to flow to affluent groups through existing spatial patterns and power, while rapidly growing informal settlements with dense housing and limited vegetation rely on the UGI that exists as a means of survival. Nonetheless, the ecosystem services that are accessed from UGI in informal settlements are of great provisioning and cultural importance to informal settlement residents. Low-income groups rely on UGI for livelihoods and resilience in urban areas,[2] making threats to UGI through LULCC, densification, overuse, and degradation a greater threat to community vulnerability. Moreover, poor UGI can act as a detractor, sometimes contributing to maladaptation through ecosystem disservices.[3]

Ultimately, the distribution of heat burdens, and of UGI's thermal regulatory services is deeply tied to considerations of environmental justice (Titz and Chiotha 2019). There is work being done by communities to take matters into their own hands, for example through local greening initiatives within informal settlements, but without formal support it will be insufficient for building adequate resilience and protecting human health (Adegun 2019). The individuals bearing the impacts of the UHI in SSA cities, and the climate crisis as a whole, are not the same individuals who created either situation.

---

studied had very different colonial and post-colonial building practices. Since most scholarship on NBS and UGI continues to be dominated by western contexts, Cavan et al.'s findings may point to avenues for future research on the socio-natural relationship between green space, formality, and ecosystem services in SSA.

[2] Contrary to the assumption that the urban poor are less dependent than the rural poor on ecosystem services discussed by Nassary et al. (2022).

[3] Ecosystem disservices are defined by Wijesinghe and Thorn (2021: 9) as "functions of natural ecosystems that are perceived to negatively affect human wellbeing" and can include vegetation being used to conceal criminal activity and outbreaks of water- and vector-borne diseases from stagnant water or open defecation in riverbeds.

Part II
Renewable Energy Development and Resilience

Chapter 8
Introduction to Renewable Energy in Sub-Saharan Africa

8.1 Overview of Renewable Energy Sources

Electricity is a modern source of energy that is absent in many parts of the world, especially in Sub-Saharan Africa where more than two thirds of the population are living without electricity. Cameroon, which is the focus of this case study, also has a low electricity access rate; 56.8% of its total population according to World Bank statistics. To mitigate this energy poverty approximately 70% of Cameroonians continue to use traditional biofuel energy for cooking, heating, and lighting such as fuelwood and coal. Biofuels are a type of renewable energy (RE) however, traditional energy use has detrimental health and environmental consequences especially in face of climate change, which creates questions on their continued sustainability. Furthermore, although Cameroon is rich in other renewable energy resources such as hydro energy and holds unharnessed solar potential, the country has failed to increase modern energy access to its population. These problems with RE development in Cameroon highlight that more concern should be accorded to sustainable energy development in order to maximize the benefits and decrease the environmental and social costs of energy development from both conventional (oil and gas) as well as renewable energy sources.

8.2 Socio-Technical and Socio-Ecological Perspectives

Two different theories, socio-technical and socio-ecological, provide alternative explanations for Cameroon's low capacity to meet the modern energy access challenge. All these theories are complex system approaches; however, they differ in what they consider to be the most important aspects of how beneficial change can be implemented and sustained. For socio-technical theory, choices, preferences, behaviors, perceptions, trust, costs, and risks all influence successful adoption and dissemination

© The Author(s), under exclusive license to Springer Nature Singapore Pte Ltd. 2025

R. Perrault et al., *Sub-Saharan African Sustainable Development*, City Development: Issues and Best Practices, https://doi.org/10.1007/978-981-95-1458-8_8

of renewable energy technologies. Socio-ecology theories highlight that local abilities, needs, traditional knowledge, adaptive governance systems, and resiliency are essential components of sustainable energy development. Sociotechnical theories emphasize the relationship between technology and its infrastructure, which "co-evolves with socio-economic institutions, actors and social norms (Goldthau 2014, 134)." The idea of sociotechnical systems was first proposed by Trist and Emery (1965) in their research on coal mines in England. An important aspect of this theory is social innovation or the social shaping of technology, which is a change in behavior that comes with the introduction of new technology. Changes in behavior can often be promoted by system changes of governance, laws, and regulations on both sides of supply and demand processes. Transitioning to renewable energy technologies is a major sociotechnical change where political and cultural factors need to be taken into consideration for consequent knowledge systems, norms, and behavior to shift (Stirling 2014). However, these complex social systems are difficult to influence, and the status quo may produce inertia against change. For example, in Japan, after the Fukushima nuclear disaster in 2011, researchers found that even after such an event, a move from a centralized nuclear energy system to a decentralized renewable energy system was not occurring (Mclellan et al. 2016). Cameroon could be experiencing socio-technical barriers to the widespread adoption of renewable energy technologies such as solar energy and mini-hydro.

The perception of renewable energies in the form of social acceptance is a key factor in the successful implementation of renewable energy projects according to many authors (Devine-Wright 2009; Wüstenhagen et al. 2007; Mallett 2007). Perceptions can be defined as a spectrum of evaluations that encompass views or feelings and that are influenced by culture, knowledge, and beliefs. Perceptions matter in the realm of RE because RE projects are smaller, more visual, and closer to communities than conventional forms of energy that are far away from city limits (Wüstenhagen et al. 2007). Where public knowledge of RE projects and their effects is low, local resistance can sometimes take place in the form of the "not in my backyard" symptom or "NIMBYism." In numerous studies, insufficient public consultation has been shown to increase the likelihood of conflict between project implementers and the public (Amigun et al. 2011; Haggett 2011; Devine-Wright 2009). The reasons for this are that projects implemented in the public space affect the locality in rarely solely positive ways. Trust can also influence public consent or dissent. Another study found that when trust was high, there was little to no public engagement in the issue, in contrast to when distrust was prevalent, engagement was higher (Smith et al. 2013). This means that opposition to the project is more likely to occur when trust is low, which could have the potential to halt or impede RE projects. Outsiders are especially perceived as untrustworthy, where private contractors instead of the community are engaged with RE development.

Another type of system approach called socio-ecology reinforces the importance of nature in shaping human society. Socio-ecological systems are complex adaptive systems composed of human and non-human entities that interact within networks (Levin et al. 2013). The similarities between socio-technical and socio-ecological frameworks are that they both "use multi-scale perspectives to examine complex and dynamic systems and recognize the principles of adaptability, transformability, and the importance of continuous learning and knowledge within the system to allow these processes (Hodbod and Adger 2014)." According to Smith and Stirling (2007), they differ in their objectives, views of progress, and how each theory frames root problems. Furthermore, ecological systems research has spatial dimensions and is subjective, while socio-technical systems are non-spatial and normative, which means that the desirability of promoting transitions and transformations is a given without questioning the unique views of different sub-groups and individuals (Smith and Stirling 2007). Social–ecological systems are more neutral on desirability by focusing on ecosystem services and how detrimental ecosystem change affects these services. Cultural values differ for each person (Hodbod and Adger 2014). In socio-ecological systems, individual behaviors are interdependent as they adapt to change in their social and ecological environment. Responding to change is an important aspect of socio-ecology, as ecological systems are never static but constantly undergoing transitions. When ecological systems change drastically, such as in the case of climate change and other instances of environmental degradation, essential environmental services such as water and soil quantities and qualities deteriorate, which jeopardizes livelihoods.

To mitigate these drastic changes, the method of resiliency has been promoted in the developmental field. Resiliency is the ability of a population to respond to change in a positive way while facing barriers, stressors, and outside forces beyond their control (Berkes 2009; Coaffee 2008). Adaptive governance or adaptive co-management are types of methods that can be utilized to increase resiliency by mitigating environmental change from a local, regional, and global level (Gaudreau and Cao 2015). This approach recognizes local or traditional knowledge, which together with scientific knowledge improves the knowledge base of development planning (Baird et al. 2005; Colding and Folke 2000). Local knowledge can also be known as lay knowledge, and traditional knowledge can also be called Indigenous knowledge, subsistence knowledge, or ecological knowledge. Examples of these types of knowledge include understanding the properties of plants, animal migrations, and other patterns in the environment that were essential to know and understand for survival. Local traditional knowledge (LTK) is defined as a specific knowledge process and way to see the world by a local group of people adapted to their environment and tested throughout generations, which has allowed them to survive, thrive, and develop their own identities and culture (Blaikie et al. 1997; Berkes et al. 2000; Taylor and de Loë 2012). A lack of community participation results in LTK being unheard or ignored, which is detrimental to development projects as proposed solutions often fail to understand the local context, needs, and capabilities.

Energy is a type of resource, and many authors argue that social-ecology is a pertinent framework that can be utilized to better pursue resilient energy development (O'Brien, O'Keefe, Rose 2007; Hodbod and Adger 2014). O'Brien and Hope (2010) define a resilient energy system as being "based on Indigenous renewable resources, capacity enhancing, adaptable and upgradable, and easy to repair and maintain (O'Brien and Hope 2010, 7556)." They argue that human development could be optimized if energy sources were locally owned and managed. Cameroon could also be facing difficulties with effectively managing their own renewable energy system and responding to change such as rapid urbanization, climate change, and environmental deterioration, which is negatively affecting socio-economic development.

Chapter 9
Case Study: Cameroon

9.1 Historical Context of Energy Development

Cameroon has the fourth highest energy potential in Africa; however, despite this energy wealth, Cameroon's energy consumption is below average for the region (Wandji 2013). It is estimated that 80% of the population does not have access to the national electric grid (Wirba et al. 2015). As electricity levels are low, traditional energy in the form of biomass holds the largest share of energy consumption. There is an energy shortage in Cameroon because of limited energy supply and increased energy demand (Adom 2016). Energy demand was 1455 MW in 2014 (Muh 2017) and was projected to increase by 5000 MW in 2020 due to population growth, putting even more pressure on the energy system (Mas'ud et al. 2015). The Cameroon government planned to implement projects to increase energy supply by 2500 MW between 2012 and 2020, but these plans have not been fully successful (Ayompe and Duffy 2014). Cameroon becoming an emerging economy by 2035 is jeopardized by the precariousness of its energy sector, which negatively affects economic growth.

Oil exportation in Cameroon began in 1977, contributing to high national economic growth rates due to a boom in oil prices and demand worldwide during the period between 1979 and 1985 (Djiofack and Omgba 2011). Cameroon's current oil reserves have been diminishing since 1985, with estimates in 2010 of total reserves of 30 million tons (Tamba 2017). However, Cameroon remains the sixth-largest crude oil producer in Sub-Saharan Africa, with an average of 76,000 barrels of crude oil a day according to the United States Energy Information Administration (USEIA). Cameroon's gas reserves are also moderate at 4.8 tcf and are globally ranked 47th on the list of the world's largest gas reserves (EIA 2018). An important aspect of oil and gas development in Cameroon is the Cameroon-Chad pipeline, the largest foreign investment project in the country. The pipeline transfers oil reserves from Chad through Cameroon, which has access to seaports. Oil is mainly utilized in the transportation industry; however, electricity from thermal gas plants powered by oil and gas supplements Cameroon's hydroelectricity. Thermal plants include Limbe,

© The Author(s), under exclusive license to Springer Nature Singapore Pte Ltd. 2025
R. Perrault et al., *Sub-Saharan African Sustainable Development*, City Development: Issues and Best Practices, https://doi.org/10.1007/978-981-95-1458-8_9

Yassa, and Kribi, among others. Besides electrical production, kerosene (propane gas) is also used for cooking and lighting.

Hydro-electric development in Cameroon began in 1953 with the completion of the Edéa power station, which has a capacity of 263 MW. The majority of the hydro dams are located on the Sanaga river basin, which is the main hydropower source of Cameroon. Both Edéa and Song Loulou (388 MW) are on the Sanaga, producing 97% of Cameroon's hydropower supply. Lagdo (72 MW) is located in the north of Cameroon near Garoua. Hydroelectricity is responsible for 73% of all electricity produced in Cameroon (Muh et al. 2018). However, Cameroon's hydro-electric system becomes vulnerable to blackouts during the dry seasons when water levels are low. Electricity rates range from 65–88% in urban areas, while the rural hinterland remains at low access levels of 14%. There are three electrical grid networks in Cameroon: the Northern Interconnected Grid (NIG), the Southern Interconnected Grid (SIG), and the Eastern Isolated Grid (EIG). Each grid supplies electricity to its respective geographic region in its name, with the NIG supplying electricity to the North of Cameroon, SIG to the South, and EIG to the East. Cameroon has the second-largest hydro energy potential in Sub-Saharan Africa after the Democratic Republic of the Congo; however, only 5% of the hydropower potential of Cameroon is harnessed (Muh et al. 2018).

The state-owned company, Société Nationale d'Electricité du Cameroun (SONEL), which originally controlled hydro-electrical development, turned to American ownership under the company AES in 2001. This privatization raised concerns among some academics because it caused a monopoly to form with insufficient regulation to ensure that affordable and equitable electrical access would occur (Pineau 2002; Imam et al. 2018). In 2014, AES-SONEL was bought by a British investment firm and renamed Eneo, which is directly responsible for managing the electrical system (production, transmission, and distribution) of Cameroon. Privatization under foreign ownership has failed to increase access levels to electricity in Cameroon, and current electricity supply is unable to meet the electricity demand of the population, especially during peak periods. In addition, large rural areas of Cameroon are not covered by the central hydropower lines, increasing the urban–rural divide. Hydro production is a key facet of future plans for economic development for the government of Cameroon, with numerous dams under construction or in the planning stages. One such hydro development, Lom Pangar with a capacity of 30 MW, is supported and partially funded by the World Bank, which argues that the increased electrical generation would benefit economic development in Cameroon (World Bank 2015). However, in 2005 a report from a local civil society organization, Global Village, and a non-governmental organization (NGO), International Rivers, highlighted the risks involved with this dam, which included flooding of forest ecosystems as well as unknown consequences on the Sanaga River, increasing Cameroon's vulnerability to droughts (International Rivers 2005). Other hydro development projects in progress include Natchigal and Menve'ele dam, as well as the rehabilitation of the Song-Loulou and Edéa power stations to increase electrical power supply to Cameroon. Cameroon also has significant potential for mini-hydro (estimated at 1.115 TWh), mainly in the western and eastern regions. However, these small hydro potentials are poorly utilized.

The majority of biomass energy consumption in Cameroon constitutes traditional energy in the form of fuelwood or animal matter. As depicted in Fig. 9.1, biomass energy holds the largest share of final energy consumption in Cameroon and has reached consumption levels similar to the previous decade. Fuelwood is cut from trees taken in areas such as the Congo Basin, which runs through southern Cameroon and is the fourth-largest tropical forest in the world (Greenpeace 2018). However, as half of the Congo Basin is under commercial logging leases, the extent of the forest ecosystem is at risk. Cameroon holds the third-largest biomass potential in Sub-Saharan Africa, as three-quarters of its territory is covered in 25 million hectares of forest (Mas'ud et al. 2015). Furthermore, as approximately 70% of Cameroon's workforce is in the agricultural sector, substantial agricultural residues could be used for biofuel production (Molua and Molua 2007). Biofuel production in Cameroon has been viable since 2005 from palm oil cultivation. The biofuel sector is predominately controlled by foreign-owned companies, which has negatively affected local communities. According to one researcher, land grabbing for palm oil plantations has occurred with little to no community consultation of the land directly implicated, and these deals often occur in secret between government officials and foreign-owned companies (Ndi 2017). Biomass energy in the form of fuelwood is largely unregulated in Cameroon, with extensive illegal logging further hindering sustainable energy development and leading to deforestation.

The solar energy potential in Cameroon is vast. However, this potential has not been exploited (Ayompe and Duffy 2014; Mbi 2013; Abanda 2012). Solar energy could provide an effective way to meet the lag in hydro-electric generation in dry seasons when the sun's rays are more powerful. The current application of solar energy in Cameroon is to power the telecommunication network, as well as other

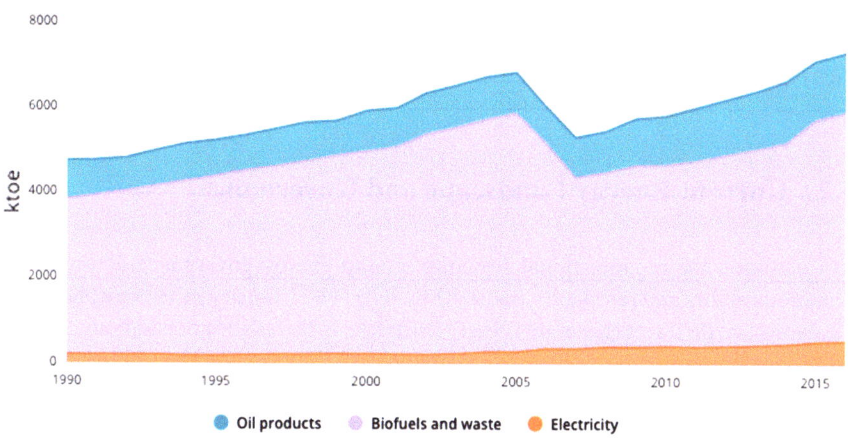

Fig. 9.1 Cameroon's total final consumption (TFC) by Source 1990–2016. *Source*: International Energy Association (2016)

small pilot projects. For example, there have been recent developments of city solar lighting installments in the major urban centers of Yaoundé and Douala (Muh et al. 2018). There are many barriers to effective solar energy implementation as current installations are not well maintained, require high start-up costs, and furthermore, there has been little government commitment to boost the sector, leading to a lack of local and international investors. Recent large-scale solar energy development projects ranging from 100 to 500 MW are in the planning stages, including JCM Greenquest, Joule Africa, and GSC Energy. The newly named company Eneo has also entered the solar market with the opening of the Djoum solar-thermal "hybrid" station, using both solar electricity and thermal electricity (Enerray 2018).

The government of Cameroon is not actively promoting solar energy, as policies are limited, such as the cancellation of the value-added tax on solar energy technology. More subsidies are required to financially support the sector. In terms of international projects, the phenomenon of donor fatigue has resulted in solar projects that ultimately failed due to poor maintenance and local management. For example, an ignorance of local needs and capacity building during program implementation negatively affected project outcomes in remote villages in Cameroon (Nfah and Ngundam 2009). Foreign companies are also starting to become involved in solar energy production. In Garoua, which is the capital of North Cameroon, a solar photovoltaic project contracted by the Italian company Enerray is being constructed with a capacity of 30 MW (UniAfrica 2017). Another solar electricity system opened in Nsem, funded by Chinese finance, in 2018 (Funwie 2018). Local small-medium enterprises (SMEs) are lagging in the solar energy sector; however, one business started by a young entrepreneur, Bolivie Wakam, called Africa Solar Tech, is making some headway (Journal du Cameroun 2017).

Wind energy potential varies throughout the country, with Northern Cameroon having higher wind levels compared to the rest of the country. However, for large-scale projects, the wind speed is not high enough to maintain energy levels (Abanda 2012). Geothermal energy potential is not well known in Cameroon, as few studies have been undertaken. The potential for tidal energy, which is energy derived from the rise and fall of sea levels in Cameroon, is also not well known or studied.

9.2 Current Energy Landscape and Governance

In Cameroon the organizational structure around energy governance comprises of numerous actors and regulatory bodies (Fig. 9.2). The Ministry of Energy and Water "le Ministère de L'Energie et de L'Eau" (MINEE) is the governmental body charged with developing energy policy and regulations. MINEE cites electricity access in Cameroon as only 28.5% nation-wide and that this negatively effects Cameroon's development in sectors such as housing, education, health, and the development of small-medium enterprises (SMEs), which is detrimental to social and economic development of the country as a whole (GTNM 2018). As a national policy goal, the Cameroon government since 2006 has been working on a multi-sectoral approach

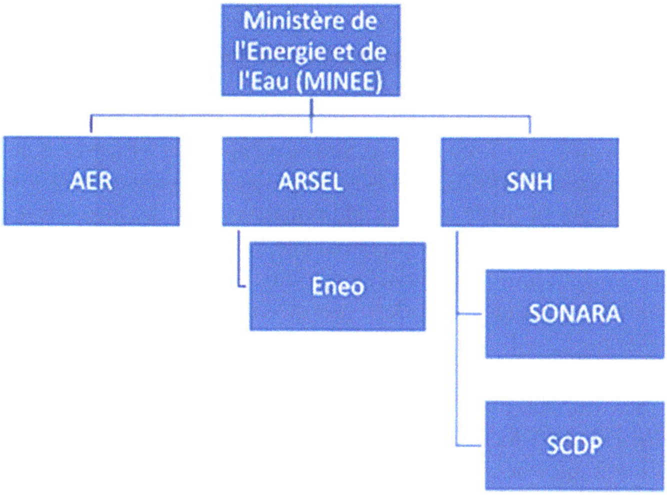

Fig. 9.2 Hierarchy of energy sector regulation in Cameroon. *Source*: Adapted from Energy Sector Management Assistance Program (2007)

called the "Groupe de Travail National Multisectoral (GTNM)" which aims to share data and increase capacity-building of governmental actors to work towards continued electrification, especially in rural areas.

L'Agence de régulation du secteur de l'électricité (ARSEL), created by law in 1988, is responsible for regulating and monitoring the electricity sector by setting electricity rates and determining electrical standards. Another agency, named the l'Agence d'électrification rurale (AER), specifically concentrates its resources on the improvement of electricity sources in rural areas. Finally, the Electricity Development Corporation (EDC) was created in 2006 to promote electricity development in Cameroon.

Electricity tariffs are established at the national level by a joint decision of the Ministry of Energy and the Ministry of Trade in agreement with Eneo and ARSEL. Additionally, three institutions are specifically involved with the regulation and implementation of oil and gas. The Société nationale des hydrocarbures (SNH) implements all the research and development in the sector on behalf of the state. The Société nationale de raffinage (SONARA) is responsible for providing the national market with refined petroleum. La Société camerounaise des dépôts pétroliers (SCDP) stocks petroleum products in distribution centers across Cameroon. Finally, la Caisse de stabilisation des prixs hydrocarbures (CSPH) was created in 1974 with the unique mandate to stabilize the price for the distribution of refined petroleum products. Eneo has an agreement with the Cameroon government to be the sole provider of producing and distributing electricity and is therefore a monopoly. On October 8th, 2018, an agreement signed in Douala by the Ministère de l'Emploi et de la Formation professionnelle (MINEFOP) established that Eneo would become a major gas supplier for Cameroon. With this agreement, Eneo is now diversifying into gas, where it already controls hydro-power in Cameroon. There is currently no institution that exists to regulate the renewable energy sector in general, which hinders growth in this sector (Tamba 2017).

The policies concerning energy development in Cameroon are numerous. Cameroon's Growth and Employment Strategy Paper (GESP), which concerns the ten-year period from 2010–2020, was heavily concentrated on energy for development, with a target of 3000 MW of additional installed hydropower capacity set for 2020. This involves the development of new power generation facilities (mainly hydro), modernization of the national electricity network, including infrastructure, and the promotion of modern and clean energy services, especially for rural communities. Furthermore, an Energy Sector Development Plan 2030 was developed with the goal of increasing electricity access to 75% total and achieving a 20% rural electrification rate. Another plan, named the Rural Electrification Master Plan (REMP), aims to electrify 660 localities with grid extensions and mini-hydro, through a diversification framework.

The link between energy and poverty was recognized in 2007 in a National Energy Action Plan for Poverty Reduction (NEAPRP), which had as its main goal to increase access to modern energy services for social infrastructure such as education and health facilities. The plan highlights energy poverty as a detrimental factor to human development in Cameroon. Throughout the report, the importance of electricity in lighting, communications, and technical services, as well as more efficient and safe cooking technology, such as "les foyers améliorés" (translated as improved oven), was promoted. Specifically, the framework of the Millennium Development Goals (MDGs) was used throughout the document to highlight the positive linkages between improved access to energy services and human development. In Cameroon, the reaction to the MDGs was the creation of the Document de stratégie de réduction de la pauvreté (DSRP), which is the national poverty reduction plan; however, the strategic plan failed to identify and utilize energy as a significant factor in reducing poverty and increasing human welfare. According to NEAPRP, replacing traditional energy with modern energy for cooking purposes is both a public safety, health, and women empowerment goal.

Cameroon is rich in energy resources; therefore, the deficit in energy supply to meet energy demand cannot be attributed to a lack of energy resource potential. Poor management, regulation, and promotion of the energy sector by the Cameroon government are the main issues at stake, according to many experts. Cameroon's electricity needs are mainly covered by hydropower. This reliance on a single energy source gives rise to Cameroon's energy insecurity and the observed regular power shortages, particularly in remote and semi-urban areas. Insufficient access to modern energy, such as electricity, results in the continued reliance on traditional energy use, which leads to problems that will be covered in subsequent chapters. Sustainable renewable energy development could be a major factor in fueling development in Cameroon. However, access to modern energy remains an issue that affects economic and socio-economic growth and hinders development in Cameroon. Increasing energy access levels while transitioning to a low carbon energy infrastructure are twin conflicting challenges. Cameroon, unfortunately, has a deficit of governance coordination and action, which hinders sustainable renewable energy development. Off-grid solutions to improve resilience and accessibility, as well as government regulation to promote renewable energy development, could meet the energy deficit of Cameroon's population.

Chapter 10
Impact and Implications of Renewable Energy

10.1 Socio-Ecological Effects

Hydro and biofuel development in Cameroon have both caused detrimental effects to the socio-ecological environment. Furthermore, health, gender, as well as socio-economic impacts on development in terms of education are all negatively impacted from the continued dependence on traditional bioenergy such as fuelwood. The unsustainable development of renewables in Cameroon affects food and water security further compounding issues of vulnerability as well as precariousness. In addition, expanding power differentials in Cameroon and the consequent unequal distribution of benefits and costs of renewable energy development results in many forms of injustices that negatively affect the well-being of local communities as well as the livelihoods that they depend on.

Environmental degradation occurs as a result of both conventional energy development and renewable energy development. Large-scale exploitative projects cause negative externalities, and if these are not properly mitigated, severe consequences can occur to socio-ecological systems. Forest cover in Cameroon was once 280 million hectares in 1965; however, biomass exploitation for fuelwood has contributed significantly to deforestation in Cameroon. According to the Food and Agricultural Organization (FAO) of the U.N., the deforestation rate in Cameroon has rapidly increased, rising from a 0.6% loss rate of overall forest cover per year from 1980 to 1995 to 0.9% between 1990 and 2010. Additionally, Cameroon lost 1.08 million hectares of forest between 2001 and 2017, equivalent to a 3.4% decrease in forest cover since 2000, which amounts to 137 million tons of carbon emissions (Global Forest Watch 2017). As of 2016, approximately 19 million hectares of forest cover remained in Cameroon (FAO 2016) (Fig. 10.1).

Illegal logging accounts for a large segment of deforestation, making it difficult to effectively manage forest resources and assess their quality and quantity. The European Union (EU) recognizes the threat of illegal logging to protected areas, and to mitigate this, a forest partnership was created to stop the sale of illegal timber. This

© The Author(s), under exclusive license to Springer Nature Singapore Pte Ltd. 2025
R. Perrault et al., *Sub-Saharan African Sustainable Development*, City Development: Issues and Best Practices, https://doi.org/10.1007/978-981-95-1458-8_10

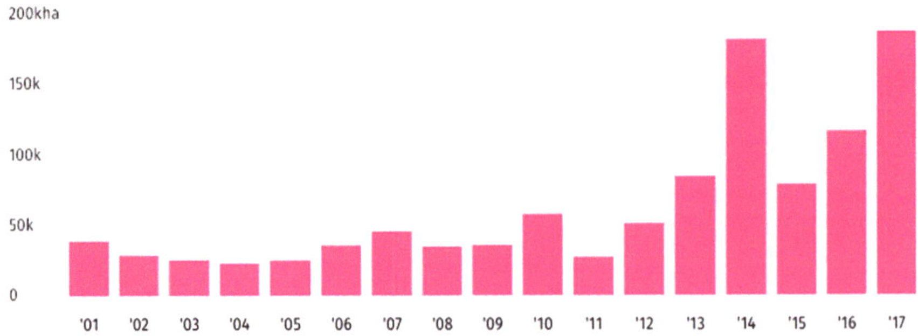

Fig. 10.1 Tree cover loss in Cameroon from 2001 to 2017. *Source*: Global Forest Watch (2017)

agreement is part of a Forest Law Enforcement Governance and Trade (FLEGT) program, which the government of Cameroon signed with the EU in 2010 (African Union 2010). The Congo Basin is an important tropical ecosystem that supports approximately 10,000 different kinds of plants and more than 2,000 different animal species, including fish, birds, and mammals. Cameroon has the second-highest rate of deforestation in the Congo Basin after the Democratic Republic of the Congo (Greenpeace 2018). Forests act as carbon sinks, reducing carbon emissions by converting carbon dioxide into oxygen, thereby supporting a healthy atmosphere. Tropical forests are also important ecosystems for many species, and deforestation threatens biodiversity. Cameroon is home to numerous species at risk, including the critically endangered western lowland gorilla and the chimpanzee (African Wildlife Foundation 2017).

Biodiesel cultivation of palm oil has also caused deforestation in Cameroon as more agricultural land is required to grow crops (Gbetnkom 2005). Continued deforestation for fuelwood and palm oil agricultural expansion exacerbates problems already occurring as a result of climate change, which is beginning to impact almost all of Cameroon's sectors of development (Bele et al. 2011). The forestry sector accounts for 6–10% of Cameroon's GDP; therefore, deforestation without the equivalent tree planting to maintain the forest cover could lead to unsustainable consequences for the economy (Bele et al. 2013). Coastal areas will be affected by rising sea levels, and other natural disasters could become more prevalent because of climate change, such as hurricanes and monsoons, which can also negatively affect the agricultural sector by destroying crops. Deforestation also impacts livelihoods as desertification and soil erosion inhibit agricultural production and may endanger human life by increasing the likelihood of landslides.

Hydro dams also have significant effects on the socio-ecological landscape. The Lom Pangar dam development in Cameroon resulted in the flooding of the Deng Deng Forest, which holds important ecological significance as one of the last hardwood forests on the continent. Interestingly, the part of the forest that was flooded was originally avoided by the construction of the Cameroon-Chad pipeline, whose costs were increased by avoiding the Deng Deng Forest. Therefore, the dam made these costs obsolete and also increases the likelihood of oil spills as the dam affects 20 km of the pipeline already constructed (International Rivers 2005). Ecological systems of animals such as fish, birds, and plants are also affected by hydro dam construction, which has consequent effects on human livelihood strategies such as fishing and agricultural activities. For example, changes in sediment levels due to dam development, which floods land, affect water chemistry, leading to ramifications on human health by making water non-potable and increasing the risk of extinction of certain species of fish. In Cameroon, the local community situated near the Mapé dam reports that many of the fish species have disappeared due to the dam. Furthermore, the positive economic benefits of hydro development have failed to reduce poverty in the local population (Mbih et al. 2014). Alternatively, reservoirs can increase the potential for economic and leisure activities such as tourism and outdoor recreation. Agriculture may be improved through irrigation, or flooding controls can be constructed to manage the variation of water levels during the dry and wet seasons. Lastly, transportation could potentially be improved by reservoirs that are less treacherous than rivers (Tchouaha 2012). Ecosystem services that forests offer, such as clean water and air provision, as well as the maintenance of fertile soil qualities, are important thresholds in socio-ecological systems. When these factors reach certain levels, they can cause feedback loops, resulting in rapid system changes that are hard to adapt to, which can result in disaster for local communities dependent on the current system.

10.2 Water, Food Security, and Health

During a World Food Summit held by the FAO in 1996, a definition of food security was internationally recognized as follows:

"A situation that exists when all people, at all times, have physical, social, and economic access to sufficient, safe, and nutritious food that meets their dietary needs and food preferences for an active and healthy life" (FAO 2002).

A report from the FAO titled The State of Food Security and Nutrition in the World 2018 estimated that 821 million people were undernourished in 2017 (FAO 2018). Therefore, food insecurity continues to be prevalent globally and is aggravated by water scarcity, which is worsened by changes in climatic conditions. Water, food, and energy are all basic human needs. However, there are limits to their exploitation as many areas in the world are prone to water and food scarcity. The over-exploitation of natural resources can lead to the phenomenon of the tragedy of the commons, where local resources of water or agricultural land are depleted to the extent that they are no longer viable. Another economic theory is the principle of trade-offs, where choices must be made between the production of different quantities of resources. For example, deciding whether land should be used for crops and livestock for food or for biodiesel production such as palm oil (Magoni 2017). As outlined in the environmental section, logging due to fuelwood exploitation and agricultural expansion leads to deforestation. Severe deforestation causes desertification, which results in soil erosion into water or being blown away by the wind. Desertification degrades soil quality and impacts food security. Furthermore, biofuel production from palm

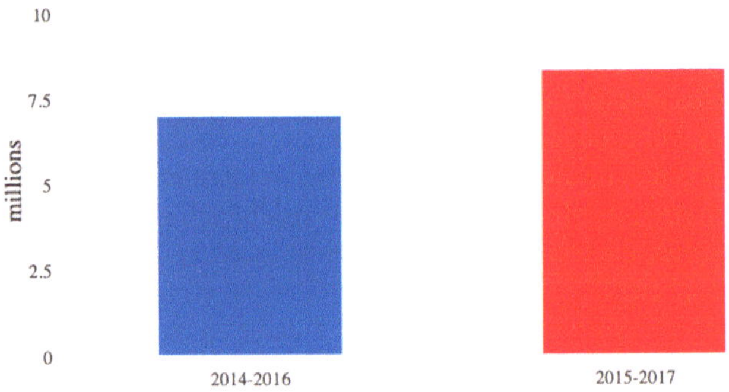

Fig. 10.2 Number of severely food insecure people in Cameroon (million) (3-year average). *Source*: Food and Agriculture Organization 2017

oil increases the run-off from fertilizers and other farming pollution, which causes groundwater salinization, negatively affecting water and soil quality and leading to food and water insecurity (Rockstrom et al. 2009). Figure 10.2 illustrates how the population that was severely food insecure in Cameroon is worsening, increasing from 7 million in the three-year period between 2014–2015 to 8.3 million people from 2015–2017.

Hydro-energy production can also affect the quantity and quality of water and food sources because rising water levels caused by hydro dams result in water salination and less agricultural land for crops. Climate change has also negatively impacted food and water security in Cameroon by increasing weather pattern variability, causing crop yields to become more unpredictable and even fail in some years (Molua 2002). With approximately 70% of the workforce employed in agriculture, this leaves Cameroon especially vulnerable to volatile and unpredictable weather patterns (Molua and Molua 2007). Periods of drought worsened by climate change can exacerbate land and water conflicts. Renewable energy has impacts on these conflicts, as well as on decisions regarding trade-offs between crops used for biofuels or food production, and water use for hydro-electricity, which can increase food and water shortages (Minucci and Karmaoui 2017). According to a UN report, there are minor implications of climate change on Cameroon's hydroelectric development as the Sanaga River is highly climate-resilient (Grijsen 2014). However, rain variations have already reduced water levels in Lake Chad. A study on water quality in the Logone Valley, located in both Cameroon and Chad, reports that the aquifers sampled contained elevated lead levels, which are dangerous for human health (Sorlini et al. 2013).

The relationship between water, food, and energy is contextualized in the Fig. 10.3, where each factor is dependent on the others, illustrating how water and food quality are impacted by energy development. Numerous types of energy production are dependent on water. Electricity is produced by heating water to generate the steam required for thermal plants, and water is also used in cooling energy systems such as nuclear energy. Conversely, potable water accessibility can be dependent on energy to pump and heat water for households, businesses, and industries. As explained previously, energy production can result in a loss of water quality due to both biofuel and hydro energy development. Water quality deterioration has consequent effects on food security due to salinization and drought. In relation to food, biomass energy is created from burning fuelwood, and biodiesel production from agricultural crops. Deforestation can result in desertification and salinization, which hinders the soil's ability to support crops, and agricultural crops used for biodiesel energy limit the food supply (A. M. Colucci 2016). Hydropower is not dependent on food; however, it results in land degradation from flooding, which negatively affects food production. Nevertheless, food security does require energy for processing and distributing food globally. For example, in this era of globalization, energy is what makes it possible to have fresh fruits and vegetables in every season. Additionally, cooking food requires energy, and modern food production is extremely dependent on energy-intensive machinery. The correlation between water and sufficient food production is apparent for proper cultivation, irrigation, and processing of food (Minucci and Karmaoui 2017).

The water, food, and energy interrelations are an example of a connected complex system where changes in one factor can affect the other, resulting in a further indirect effect on the third factor. This interconnection highlights that there may be tensions that exist between the agricultural, food, and energy sectors which consequently impacts human well-being.

The interrelations between water, food, and energy are an example of a connected complex system where changes in one factor can affect the other, resulting in a further

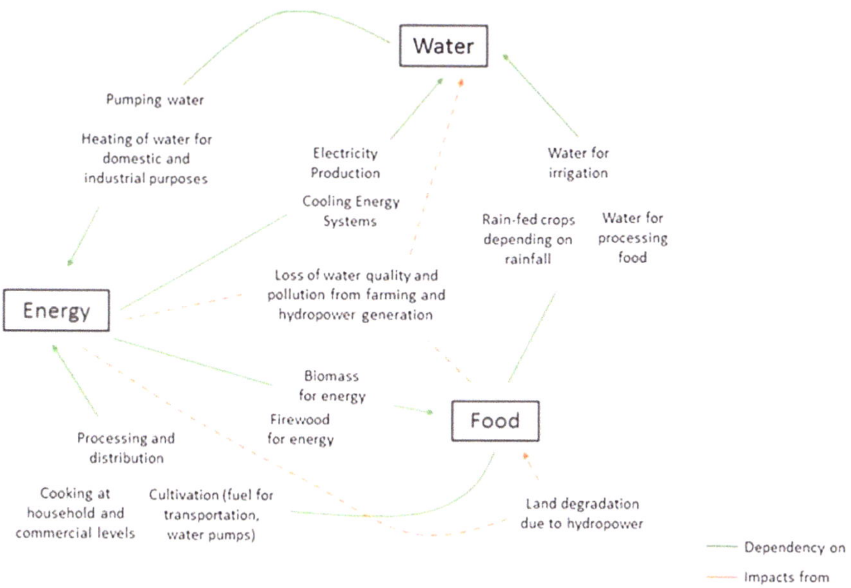

Fig. 10.3 The water, food, and energy nexus. *Source*: Minucci and Karmaoui (2017)

indirect effect on the third factor. This interconnection highlights that there may be tensions that exist between the agricultural, food, and energy sectors, consequently impacting human well-being.

Burning traditional energy sources such as fuelwood has significant consequences on health. Indoor air pollution (IAP) can result in "acute respiratory infections, tuberculosis, chronic respiratory diseases, lung cancer, cardiovascular disease, asthma, low birth weights, diseases of the eye, and adverse pregnancy outcomes" (Sovacool 2012, p. 275). The combined factors of indoor and outdoor air pollution have caused over 7 million people to live in conditions of unsafe air, according to a State of Global Air report in 2018. Breathing polluted air has contributed to 6.1 million premature deaths due to diseases of the cardiovascular system such as stroke, lung cancer, and heart attack (World Energy Assessment 2000). In terms of the global burden of disease risk factors, there are just three other forces that cause more harm globally, which are high blood pressure, diet, and smoking. In Cameroon, child pneumonia and chronic obstructive pulmonary disease (COPD) are heavily linked with solid biomass burning for cooking and heating. A UNDP report titled The Energy Access Situation in Developing Countries cited that in Cameroon, 9,900 children under 5 died from pneumonia and 1,500 adults died from COPD in 2004 (UNDP 2009). In addition to IAP, physical injury can occur during fuelwood collection, including backaches from carrying heavy loads of wood over long distances, as well as foot damage and wounds from accidents (Sagar 2005). Figure 10.4 illustrates that the annual number of premature deaths attributed to illness as a direct result of IAP from the use of solid fuels for cooking and heating in Cameroon has risen steadily since 1990. This trend contrasts with Sudan, which has reduced the number of deaths attributed to IAP. Therefore, Sudan is converging on Cameroon, while countries such as Botswana have successfully maintained low rates of deaths related to IAP that are similar to those of developed countries like Canada.

Inadequate access to modern energy services also results in a lack of refrigeration and proper lighting in health facilities, hindering proper vaccination, blood supply storage, and the ability to perform surgeries and other lifesaving health activities. Information and communication technology also aids in the dissemination of health awareness. A lack of electricity to power computers and other electronic devices can reduce the effectiveness of mass media such as television and radio to promote sanitation norms; for example, condom usage to decrease the spread of diseases such as HIV/AIDS (Sovacool 2012). Furthermore, hydro reservoirs can act as breeding grounds for vector-borne diseases such as malaria, spread by mosquitoes. Non-potable water due to hydro dams also increases the likelihood of waterborne illnesses. However, renewable energy can also offer solutions through the use of solar energy to power pumping stations that provide potable water.

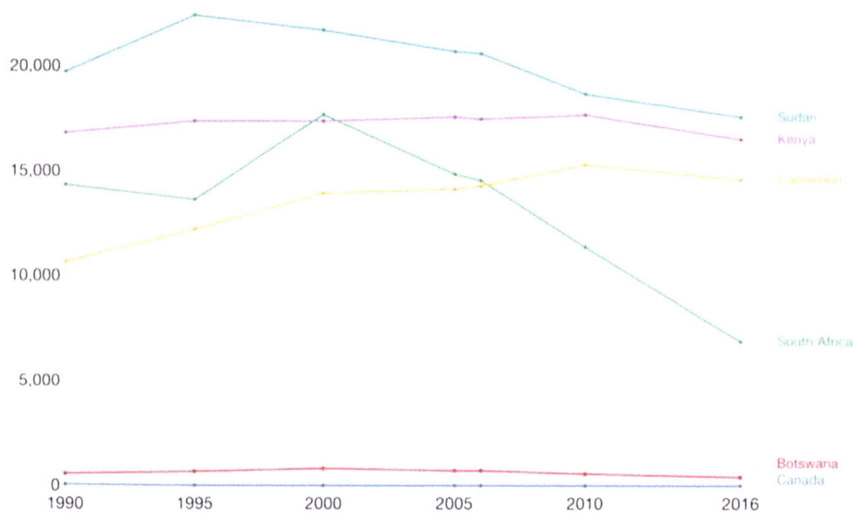

Fig. 10.4 Absolute number of deaths from indoor air pollution 1990–2016. *Source*: Roser and Ritchie (2013)

10.3 Gender, Inequality, and Urban Environment

Energy poverty is defined as a "lack of choice to adequate, reliable, affordable, efficient, safe, and environmental benign energy services to support economic and human development" (Masud et al. 2007). Energy poverty contributes to continued income poverty and the poverty cycle because the percentage of income attributed to the consumption of energy services is higher for poorer individuals. Energy services include lighting, refrigeration, and its ability to power devices and machines that have consequent effects on the provision of health and education (Karekezi and Majoro 2002). Communication and information technology are also modern energy services powered by electricity. Agriculture activities such as "water pumping for irrigation, fertilizer, mechanized tilling, and processing which increases crop yields and productivity" all require access to energy (Huang and Chen 2005) Additionally, on a household level, domestic time-saving appliances, cooling, heating, and transportation are necessary energy services of the modern twenty-first century. Limited access to modern energy such as electricity hinders the provision of lighting and the ability for students to study at night or for businesses to stay open late to increase sales and therefore, has a direct relationship with education and income levels.

Energy poverty is also highly linked with gender. Women are traditionally tasked with household activities such as cooking. As previously stated, burning fuelwood to cook food creates indoor air pollution (IAP) that negatively impacts health. Therefore, as women are closer to the fumes, their health is impacted the most. Furthermore, collecting and retrieving fuelwood are also tasks that women are responsible for in most traditional cultures. Cuts, wounds, and injuries can occur during these tasks. In addition, women face exposure to hazards such as extreme weather and an increased risk of sexual assaults, increasing the danger of these activities. Women are often forced to bring children with them, exposing them to the same risks (Masud et al. 2007). The time required for cooking and other household chores, such as collecting fuelwood, also impacts the ability of women to pursue income-generating activities that promote independence and empowerment (Sagar 2005). Furthermore, female educational levels are also negatively affected by traditional energy use due to absenteeism caused by injuries and sickness (UNDP 2009). Absenteeism affects girls more than boys because of the above-mentioned barriers to health, as well as the time needed for cooking and finding fuelwood, which are generally culturally feminine tasks. In Cameroon, female absenteeism and educational attainment are significantly lower than their male counterparts. For example, the mean years of schooling for females was 4.7, while for males it was 7.6 in 2017 (UNDP 2018). The Human Development Index for Cameroon also shows that female labor participation is ten percent lower than that of males, and, as explained, traditional energy plays a role in women's ability to perform non-household work and earn income.

Time-saving devices and appliances powered by electricity benefit women by improving the safety and efficiency of tasks such as grinding food and retrieving water. The provision of electricity improves conditions at home through services such as lighting and refrigeration and also increases the possibility of generating income through specialized small-medium enterprises (SMEs). Adequate lighting at night also improves the safety of women traveling outside in the evening, possibly reducing the likelihood of rape. Finally, electricity can power organizations that provide social amenities for women (ESMAP 2007). Electrical rates for these amenities that serve the public good are low in Cameroon.

The relationship between modern energy services and development is substantial for low-income countries due to a reduction in the necessity of biomass energy sources, which has detrimental effects on health and gender, as already discussed (Martínez and Ebenhack 2008). There are also strong links between energy and urban development according to many authors, which positively contributes to economic growth (Huang and Chen 2005; Rutter and Keirstead 2012; Madlener and Sunak 2011). The main argument for this correlation is that the urban environment requires more energy than the rural hinterland as a center of economic growth, innovation, and as a hub of transportation and massive infrastructure investments (Madlener and Sunak 2011). Global energy demand, therefore, should increase as urbanization continues, with rural-to-urban migration leading to a concentration of people in cities and a structural change in the economy as agriculturalists become laborers that produce goods and services. This change further requires a transition from low-intensity to high-intensity production, resulting in more urban energy demand.

The buildings and infrastructure required for an urban environment are also energy-intensive, especially with the necessity of heating and cooling due to seasonal changes in temperature. Energy is also necessary for transportation and trade to move goods, services, and people in and out of urban centers. Furthermore, as fewer people are agriculturalists because urbanization results in a decreasing rural population, the farming sector becomes increasingly mechanized, requiring additional energy (Huang and Chen 2005). The process of urbanization has been the global trend, with 55% of the world's population living in cities and projected to reach 68% by 2050 (UN 2018). Africa is urbanizing at a faster rate than ever before, and Cameroon is no exception, with its urban population increasing from 13.94% in the 1960s to 55.78% in 2017 (World Bank 2017). Global energy demand, therefore, should increase as urbanization continues, with rural-to-urban migration leading to a concentration of people in cities and a structural change in the economy as agriculturalists become laborers that produce goods and services. This change further requires a transition from low-intensity to high-intensity production, resulting in more urban energy demand (Madlener and Sunak 2011). The buildings and infrastructure required for an urban environment are also energy-intensive, especially with the necessity of heating and cooling due to seasonal changes in temperature. Energy is also necessary for transportation and trade to move goods, services, and people to and from urban centers. Furthermore, as fewer people are agriculturalists because urbanization results in a decreasing rural population, the farming sector becomes increasingly mechanized, requiring additional energy (Huang and Chen 2005).

Urban energy systems are defined as the "sectors and processes needed to satisfy the energy service demands of an urban area" (Walker et al. 2015). The world is becoming more urban, so the future of urban energy systems will be a key determinant of meeting energy demand in ways that are sustainable (Rutter and Keirstead 2012). Figure 10.5 illustrates that the share of primary energy demand is projected to change drastically on a regional level by the year 2040. While developed countries such as the United States, Japan, and areas of Europe will experience a decrease in energy demand, developing countries such as India, China, and parts of Africa comprise the largest share of future energy demand. This augmentation will increase the pressure on energy supply, which, if not met, could inhibit economic growth and cause political unrest. With the population in Africa projected to double by 2050, 80% of this growth will likely occur in cities, mounting pressure on municipal governance to provide for the rising energy demands of rapid urbanization (World Economic Forum 2018).

In many developing countries, the positive relationship between energy and urban development seems to be becoming increasingly unclear. Rapid urbanization is leading to urban deterioration as poor urban planning, insufficient fiscal resources, and environmental degradation decrease urban dwellers' abilities to improve their quality of life (Nyambod 2010). The reasons for this are that while urbanization and economic development were in the past mutually reinforcing, urbanization is now occurring in developing countries without equivalent economic development. Problems arise, such as the phenomenon of poorly planned cities, which further hinders energy efficiency. For example, unplanned cities account for congestion due to individualized transport, which negatively affects the urban environment due to emissions. Pollution causes particulate matter to form in the air—tiny particles that, over the long term, can cause breathing problems in humans when their concentration reaches certain levels. A 2014 study listed three cities in Cameroon—Bafoussam, Bamenda, and Yaoundé—that experience significantly higher levels of particulate matter than those recommended by the World Health Organization (Antonel and Chowdhury 2014).

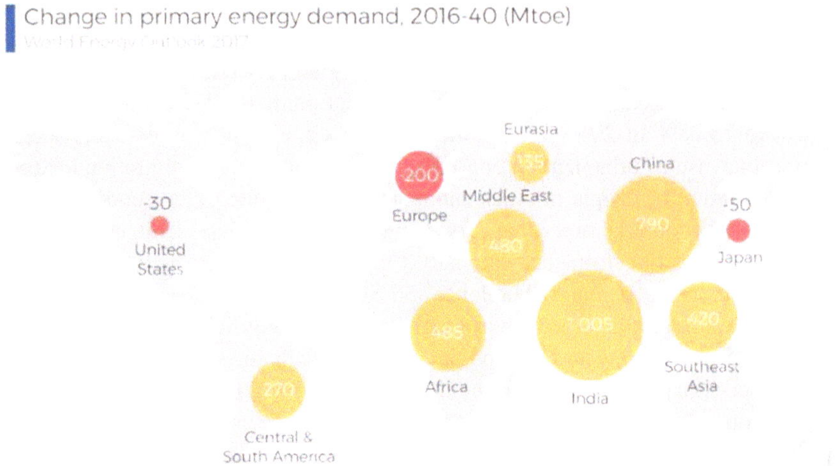

Fig. 10.5 Projected change in primary energy demand from 2016 to 2040 (Mtoe[19]). *Source*: IEA (2017)

Urbanization requires space to house a rapidly expanding population which, when badly planned and implemented, can hinder effective energy development. For example, informal settlements such as slums often lack energy infrastructure and instead rely on traditional sources of energy for cooking and heating, which also causes health concerns such as IAP (UNDP 2009). Illegal hookups into the electrical grid can result in electricity shortages, increasing the costs of providing electricity and posing a safety hazard. An estimated 60% of Cameroon's urban population lives in slums and informal settlements (UN-Habitat 2014). The World Bank reported that 37.8% of the total Cameroon population lived in informal settlements in 2014. Additionally, ineffective energy infrastructure directly impacts economic development as a whole. The lack of proper lighting and frequent blackouts negatively affect business, lowering foreign direct investment (FDI) due to higher risks, costs, and smaller returns of doing business in Cameroon. The unreliable electricity system in Cameroon is estimated to cost the country 2% of GDP every year (Kenfack et al. 2011). According to an African Infrastructure Diagnostic Report (AIDR) published by a division of the World Bank in 2011, Cameroon is facing major challenges in meeting basic infrastructure needs, including access to modern energy services, which impedes development (Dominguez-Torres and Foster 2011).

The three different types of injustices that can occur in renewable energy development projects when there is insufficient community consideration and consultation are outlined by Calzadilla and Mauger (2018). These injustices are distributional injustice, recognitional injustice, and procedural injustice (Calzadilla and Mauger 2018). All three types of injustices have occurred in Cameroon during renewable energy development project implementation. The first type of injustice, distributional injustice, occurs when the benefits and costs of RE projects are unequally distributed. An example of this is hydropower development, which causes problems for local communities worldwide that negatively affect livelihoods and ecosystems that communities depend on (Farfan et al. 2018). In Cameroon specifically, there have been instances reported of hydro development that has adversely impacted local communities. During the development of the Mapé Dam in North-West Cameroon, there is evidence to support that forced relocation occurred and that, in the process, important sites to the Indigenous community living in the area were lost. For example, gravesites that have cultural and historic value were flooded as well as the communities' homes. In addition, the community received insufficient compensation, and many people were forced to leave the community (Mbih et al. 2014). Where the Mapé dam illustrates the unequal costs of hydro development on local communities, another dam development, Lom Pangar, shows that benefits of hydro dam development were unequally distributed. Electricity generated from the Lom Pangar dam goes primarily to an aluminum smelter, Alucam, whose profits are owned by a Canadian company, Alcan, rather than benefiting the local people who will continue to face blackouts and power disruptions (International Rivers 2005). Electricity access rates are unequal in Cameroon, especially between urban and rural areas. There are also injustices resulting from corrupt practices to curb dissent in Cameroon. Government-controlled police forces have, on numerous occasions, targeted protestors to preserve control. For example, students protesting the lack of electricity at universities were shot at by police, resulting in three deaths in 2007. Electricity provision has also been used to control English-speaking Cameroonians, who comprise 20% of the population. Amnesty International reports that during the federal election in 2018, there were incidents of house arrests of oppositional leaders, the use of anti-riot police, as well as journalist arrests affecting free speech and freedom of the press. A targeted electricity blackout was also carried out in English-speaking regions, where internet and telephone communications were cut from January to April 2018 (Amnesty International 2018).

A second type of injustice, recognitional injustice, is the disrespect for people and cultures directly impacted by the development of renewable energy projects. In Cameroon, biofuel production has resulted in recognitional injustices where project developers often use the term "wasteland" to highlight that development is occurring and give the illusion of benefits, but this "wasteland" may have strong value for local communities. An example of recognitional injustice in southwestern Cameroon is the case of Herakles Farms, which is a palm oil plantation under SG Sustainable Oil Cameroon (SGSOC), owned by an American corporation. The local community contested the development because agricultural land was taken away from them without proper compensation or provision of alternative means of making an income (Ndi and Batterbury 2017). Despite this local protest, which was bolstered by international actors such as Greenpeace, the development continued and, as a result, destroyed the traditional lands of two communities living in the area in 2015. Local livelihoods were negatively impacted as landowners lost control of their traditional farmlands, and benefits to the community were limited as jobs that were promised never materialized, and communities still did not have access to electricity despite the close proximity to the palm oil development. Furthermore, one local activist, Nasako

Besingi, who protested the development, was wrongly convicted and sentenced to three years in jail for merely exercising his democratic rights (Greenpeace 2015). The uncertainty surrounding the development due to failed promises of providing local jobs and insufficient community consultation led to widespread distrust towards SGSOC by local communities (Ndi 2017). There is also evidence that Herakles Farm violates national laws and international conventions because illegal logging took place in 2010 without proper authorization from the Ministry of Environment to assess the environmental, social, and economic impact of the development, and local protests were met with intimidation and, in some cases, arbitrary arrests (Ashukem 2016). The differences in power between large foreign companies such as Herakles Farm and local communities are significant, as small villages lack the financial and legal capacity to take these giant firms to court and have legal compliance to protect local ownership of land (Fonjong et al. 2016). The ability of these giant companies to take over large tracts of land without the due consent of the local communities living on the land is defined as land grabbing. Other foreign-owned companies, such as Sithe Global Sustainable Oils in Cameroon, have also been exposed to similar forms of injustices towards local communities due to their failure to acquire informed consent from the local population directly impacted by the project's development (Pemunta 2018).

The final form of injustice, procedural injustice, is the lack of genuine information and public participation processes during renewable energy development planning and implementation. Community consultation is important because there are environmental costs from renewable energy development, as discussed previously, that could negatively impact health, safety, and local livelihoods. Regulatory measures to minimize these consequences, such as Environmental Impact Assessments (EIA), are often less widely regulated and enforced than in developed countries. Therefore, insufficient understanding of the consequences of renewable energy development and decisions on these trade-offs can lead to injustices in project outcomes (Calzadilla and Mauger 2018). Procedural injustices include a failure in EIA implementation when impacts of the project on the local community are unaddressed and local concerns are ignored. A lack of community consultation and consent results in feelings of violation for stakeholders, which can compound distrust and dissent towards developers. As explained, palm oil plantation corporations, such as SGSOC, often fail to consult with community members to gain their perspectives and address their concerns. In addition to biofuel development, hydro dams are often planned and implemented between government officials and foreign-owned companies behind closed doors without proper consultation with the local communities directly impacted. Illegal logging and deforestation also threaten communities in Cameroon, including Indigenous communities, by compromising traditional livelihoods such as hunting and gathering (Greenpeace 2019).

Renewable energy development in Cameroon has had significant negative impacts on social, economic, and ecological systems. The current energy system has been ineffective in increasing access levels to modern, reliable, and safe electricity in Cameroon while mitigating the negative effects of energy development on the environment and the people that rely on traditional livelihoods. Furthermore, unequal power distributions between foreign-owned conglomerates and the Cameroon government, on one hand, and the communities that are disproportionately affected by renewable energy development, is creating injustices, conflict, and violence, further inhibiting sustainable energy development.

Chapter 11
Theoretical Applications and Policy Recommendations

Traditional biofuel energy in the form fuelwood, coal and dung is primarily used in Cameroon for cooking and heating purposes. The need to transition to more safe and efficient forms of energy to improve the well-being of woman who are severely impacted from breathing IAP is an important aspect to reduce energy poverty. Although gas is not renewable and has detrimental effects to the environment, it does provide a safer alternative for cooking energy because it results in far IAP and is more efficient than burning fuelwood. However, domestic propane gas for cooking remains very expensive for the majority of the Cameroon population with user rates at less than 50% in Yaoundé (ESMAP 2007). Electrical connection to the national grid is also expensive because although electrical rates are 50 CFA Francs per Kwh, there remains substantial costs to connecting to the national grid (Eneo 2018). Therefore, the demand for electricity and gas could be inhibited by the high initial costs of transitioning to new technology such as a gas or electric stove. End-users are often forced to borrow money or establish community partnerships in order to afford new devices. Petroleum products are also expensive because they are primarily imported from other countries, which adversely effects Cameroon's energy security (Enongene et al. 2016). Tariffs on oil and gas imports further exacerbates the high costs for the poor in comparison to their wages and the cost of living.

A theory of energy development called the "energy ladder" outlines that as income increases, a transition towards more modern and efficient energy sources occurs (UNDP 2009). However, this does not seem to be the case in Cameroon, where even higher income groups continue to utilize fuelwood for cooking and heating (Hiemstra-van der Horst and Hovorka 2008). The rationale behind this juxtaposition could be a high-risk aversion to the costs required to transition to modern energy such as electricity. Low income and limited infrastructure are two barriers that prevent households from connecting to the electrical grid. Electricity requires a permanent structure which is built according to electrical codes. Building a new home or making changes to current home in order to support a connection to the electrical grid is very expensive and therefore, unfeasible for poorer households, especially in rural

© The Author(s), under exclusive license to Springer Nature Singapore Pte Ltd. 2025

R. Perrault et al., *Sub-Saharan African Sustainable Development*, City Development: Issues and Best Practices, https://doi.org/10.1007/978-981-95-1458-8_11

areas (Sovacool 2012). The initial costs of electrical grid connection are expensive, requiring an electrician and materials such as wires which is expensive. In addition, a monthly electric bill is risky as income is more precarious and unknown due to other potential costs such as health care and investment in education which may overrule electrical connection to the national grid.

Other factors such as traditions, social expectations, and cultural barriers may also be responsible for hindering the modern energy transition. Culturally, Africans prefer cooking over the fire which many consider to result in food tasting better with a smoky flavor (Murphy 2001). Another barrier that could be negatively impacting perceptions of renewable energy is that the benefits of modern energy technology may be unknown to the majority of the population. Energy literacy comprises of knowledge about energy and where it comes from, as well as recognition of alternatives such as renewables (Comeau et al. 2015). A study undertaken in South Africa illustrated that the factor that was the most influential on energy literacy was proximity, and that the closer or more familiar a RE technology or project is to the respondent, the more knowledge they seemed to have of the inner mechanism and functions of how the RE technology functions (Simanauskaite 2014). In addition to proximity, education seems to also be a factor in the domain of energy literacy according to another study that found there was a strong correlation between education on RE and energy saving behavior in students (Halder et al. 2010). Perceptions in renewable energy development are can also be shaped by the levels of trust towards project developers and decision makers from the local community. A survey from a Canadian study illustrated that Aboriginal Canadians were more distrustful in the information from the public and private sources than non-Aboriginal Canadians which could be due to their historical mistreatment by government authorities (Moore et al. 2014).

The reasons for the poor implementation of transitioning to more modern energy for cooking in Cameroon could be the high initial price of changing technology for gas and electricity as well as a government and private partnership failure in providing reliable and affordable gas and electricity access. All these factors keep the price high and demand low for propane gas and electricity so that the transition does not occur. The supply of energy services in Cameroon is also a barrier to effective energy transitions. Private enterprises are not effectively providing energy access because of a lack of competition. The stockage and replenishment of propane tanks/bottles seems to be a key issue which increases the risk and lowers the cost-savings and availability of gas as a cooking energy source (ESMAP 2007). In addition, there are high risks to entering the market as security is required to protect gas depot centers from theft. Market failures can also result from the formation of monopolies that can afford to keep prices artificially high in order to maximize profits because they are the sole provider of the good or service. In Cameroon the CSPH is solely responsible for stabilizing the price of gas, and therefore, there is a disincentive to lower the price and in turn their profits by increasing the supply and availability of gas. In addition, the formation of a monopoly can also been seen in the electrical sector of Cameroon as Eneo is the sole provider of electricity, therefore, there is a disincentive to lower prices or provide electricity for rural areas. This increases the likelihood of the phenomenon of "uneven-electrification" to occur where electric monopolies prefer to serve urban

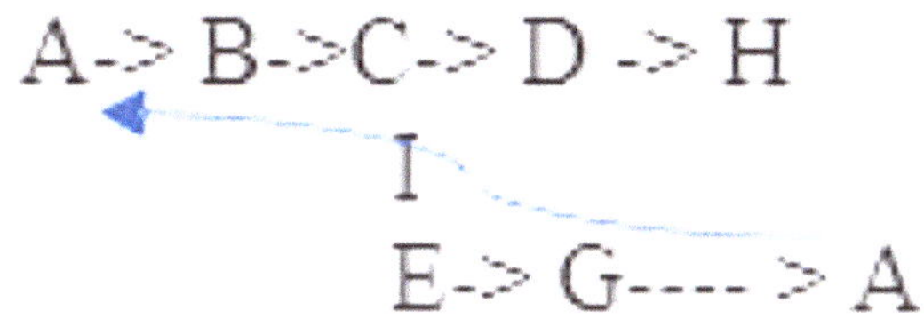

Fig. 11.1 Feedback loop illustration. *Source*: Own illustration (2019)

centers rather than rural areas because potential users and profits were more assured (Cook and Bank 2005). Connection to the electrical grid in remote areas is more costly and poorer populations lack the ability to pay service and connection fees. Therefore, the private sector is often unwilling to supply electricity to rural areas with lower population densities and poorer populations who lack sufficient income or access to credit to pay for electricity. This phenomenon can be seen in Cameroon with rural electrification access at rates far lower than urban areas.

Given the low purchasing power of the population in developing countries such as Cameroon and the economic constraints that hinders RE development, renewable energy requires an enabling environment for its promotion in the form of government policies and regulation. Socio-technical theories offer explanations for poor access rates to modern and affordable energy, but this theory offers limited rationale for other components of sustainable energy which includes renewable energy development effects on the environment and the local communities who depend on traditional livelihoods and how these factors are mitigated. These considerations are important because they impact the abilities of future generations to meet their energy needs which is an essential component of sustainable energy development.

Social–ecological systems are complex "integrated systems in which humans are part of nature and therefore cultural, political, social, economic, ecological and technological components interact (Hodbod and Adger 2014)". The theory of socio-ecology argues that socio-ecological systems are always in a state of change because of the interacting relationships between actors that cause feedback loops due to direct and indirect changes in one factor that causes change in another which when drastic can alter the entire system. For example, Fig. 11.1 shows that factor A influenced factor B, which in turn effected C, which caused changes in both factor C and factor E. Factor E caused a change in G which resulted in a feedback loop where the original factor A is impacted which results in the entire system changing yet again. Unlike socio-technical systems a perfect equilibrium can never be maintained as feedback loops occur between actors and over time with sufficient pressure, the system eventually crosses thresholds that result in regime shifts that could be irreversible and disastrous. To mitigate these constant changes in a sustainable way there are three principles of social–ecological systems to reduce vulnerabilities which are; resilience (ability to absorb shocks and disturbances), adaptation, and capacity building (Levin et al. 2013).

Anthropogenic climate change increases the necessity for humans to notice and respond to rapid ecological change in order to adapt affectively. For example, understanding weather patterns and behavior changes in animal species. In practise there is robust evidence that integration of local traditional knowledge (LTK) with additional scientific and technical knowledge can improve disaster risk reduction and climate change adaptation (van Aalst et al. 2008). Both of which improve the sustainability and resilience of communities and their livelihoods. Examples of this include indigenous knowledge of early warning signs of a disaster that saves lives by people evacuating the area. It has also been argued that LTK has been underutilized in the development field, and that this oversight can have serious repercussions on the successful implementation of development projects (Taylor and de Loë 2012; Berkes 2009; Blaikie et al. 1997; Agrawal 1995). Although ties between RE development and LTK have not been previously defined, LTK does have practical applications in other fields related to RE, including, resource management, urban development, and sustainable economic development. All of these different constructs also relate to RE because it is a natural resource that is a fundamental component of the modern built environment. A study undertaken in Buea Cameroon that focused on water supply and the effect that urbanization had on community management scheme showed the complexity of responding to social-ecological change. The researchers found that with urbanization comes increasing complications in effective management and institution building largely as a result of a rapidly expanding heterogenous population. Namely that the myriad of people in the growing community were not well represented, especially the "outsiders" coming into the community and this caused internal conflicts that stalled the project (Sally et al. 2014). A further study in another African county, Ghana illustrated that communities that were homogenous were more effective to locally manage water and sanitation services in rural environments than communities that were "ethnically diverse" (Sun et al. 2010). Urban environments are more complex and therefore acquiring united consent and ensuring that all of an urban communities' diverging needs and concerns are met proves very difficult.

Many authors argue that LTK is instrumental in sound resource management practises (Baird 2007; Taylor and de Loe 2012; Clark and Brake 2009). As resources become scarcer it becomes more important to manage these resources wisely. Baird (2007) asserts that Community Based Co-Management (CBC) is instrumental to addressing problems of a socio-ecological nature. In his paper he explains how the Local Ecological Knowledge (LEK) of the people living near the Mekong River in Southern Laos solved the tragedy of the commons taking place. The fisheries were being depleted as no one owns the fish, and therefore the livelihoods of fisherman were at risk. Using their LEK to monitor and adapt to this problem, the CBC framework was successful in allowing for decision making processes that created "regulations to meet local conditions (Baird 2007, 87)" which meant that the community respected and abided by them. In this way the fish stocks and the likelihoods of the fisherman were saved. In Cameroon, during the development of the Mapé dam, the local community noticed that the local fish stocks were negatively impacted (International Rivers 2005). Another rationale behind the LTK approach to resource management is that the people on the land have the most knowledge about their resources and have the most to lose if these resources are not managed sustainably. In multiple cases integrating scientific and local knowledge together in resource planning resulted in positive outcomes where knowledge was improved in a general sense and innovation into ways to protect valuable resources was generated (Clark and Brake 2009; Giordano et al. 2010; Kroon et al. 2009).

LTK is also applicable to sustainable urban development as well because of its ability to improve the knowledge base of urban planning which increases the health, safety, and wellbeing of urban inhabitants. For example, Corburn (2003) details how "environmental problem solving" that is linked to behaviors and the local environment has begun to be met with success in some cities in the United States, and its implementation in the developing world could have drastic benefits to global health. His examples of how missed information could have led to an air toxicity problem which otherwise would have become a dangerous health hazard in the community shows the validity of LTK in the urban context (Corburn 2003). Another study shows how Local Ecological Knowledge (LEK) similar to LTK "can complement ecological research information and indicate objects and places important to stakeholders (Yli-Pelkonen and Kohl 2005). This information is beneficial because urban planners can then preserve spaces of communal importance which protects the uniqueness of the locality by benefiting the positive psyche of citizens. Proper urban planning should take into consideration local knowledge and even traditional knowledge as it allows for specific contexts to be recognized and respected, ensures that there is proper public engagement and support, and finally brings to attention crucial information that has the tendency to be overlooked by the "professionals." This evidence supports that LTK can improve the knowledge base of both lay people and the professionals for "joint fact-finding" to occur (Mallett 2007).

Indigenous knowledge systems were also recognized to benefit the sustainability of projects in a rural study of a water provision project performed in Cameroon where the participants' local knowledge informed the projects' design and implementation (Fonjong, Emmanuel, Fonchingong 2005). Another study also performed in Cameroon illustrated the consequences of insufficient community engagement in a water project which impeded the projects' effectiveness. A failure on behalf of the developmental project's authorities to effectively take under consideration the population's local knowledge resulted in a lack of holistic understanding of the unique local needs and conditions that were implicated. Furthermore, the sustainability of the project was jeopardized because social results such as capacity building were seen as less important than the physical outputs of the project. Failing to focus on the community's strengths and insufficiently addressing their weaknesses, the resilience of the community was largely unimproved according to one research. The same researcher argued that the larger objectives of developmental projects such as poverty reduction, community participation and institution building could have been better attained if local traditional knowledge was considered and better integrated during the projects' implementation (Njoh 2003). Additional benefits of community participation include local empowerment, democratic promotion as more voices are being heard, low cost solutions that receive more support, and promotion of justice, especially of the poor by addressing local needs (Corburn 2003).

As previously illustrated, food and water security are serious issues threatening health and exacerbating energy poverty in Cameroon. LTK could enhance the transition to modern cooking technologies and involvement of local decision-making in co-management of water sheds and forests could result in improved knowledge sharing on resources which could improve sustainable energy development. In the realms of sustainable development, LTK can "benefit cooperation, stakeholder participation, understanding the long-term effects of project development and recognition of inter-generational equity and justice, as well as moderate production and consumption habits" (Eyong 2007). Additionally, LTK has the potential to foster more socially egalitarian and environmentally sustainable relationships between human societies and nature (Ellis 2005). Furthermore, sustainable farming practises that incorporate technology and innovation with the local knowledge systems may increase local capacity to enhance food security and reduce adverse effects to the environment. However, power relations may hinder the ability of local communities to participate in socio-ecological systems which creates injustices and has consequences for sustainable energy development.

11.1 Political Ecology Insights

Conventional energy development refers to top-down governance of the state for implementing electrical provision from non-renewable energy sources such as oil and gas. However, according to some authors there is a disconnect between how conventional energy systems have been organized and how end-users make decisions on household energy consumption. In the United Kingdom top-down energy policy has not resulted in desired changes in energy efficiency behaviors of the population. One research team argues that citizens do not require access to information to make informed decisions but require "community knowledge networks" where recognition of local contextual knowledge is recognized and promoted. The authors highlight the different between information and knowledge and argues that energy usage must be contextualised in how individuals actually use energy in their households. The found that community approaches increase participation, engagement, trust (Catney et al. 2013). Top-down energy development is ineffective for numerous reasons. Governments should regulate the energy sector to curtail corruption as well as implement policy to make energy transitions more feasible. However, government officials are often occupied with other sectors that generate government revenue and also lack sufficient funding and effective coordination across ministries to promote renewable electricity development, especially in rural areas. Strong governance and effective institutions are lacking in Cameroon as well as other parts of the developing world due to high levels of corruption. Another study found that the Mexican public was largely uninformed about RE technology and that there was inadequate communication between sectors of society such as the government, universities, and private enterprises that hindered the development of renewable energy development (Mallett 2007).

International NGOs that promote top-down energy development are often poor at increasing capacity building of the community to manage and maintain renewable energy project technology which has led to unsuccessful projects in the long term. A main factor behind this tendency is the traditional view of North–South relations which argues that a technology transfer should occur from the developed world to the developing world. The technology transfer between North and South is criticized by Mallett (2007) who instead promotes a "technology cooperation" model that comprises of energy literacy and energy citizenship where participants play an active role in renewable energy development (Mallett 2007). Energy literacy has been discussed previously in the socio-technical section. Energy citizenship involves public engagement on energy issues as well as increased local production and management of energy sources (Comeau et al. 2015) This concept is closely related to energy literacy because knowledge can encourage engagement with renewable energy, and reversely engagement with RE incentivizes gaining more knowledge on energy development and its impacts on communities. Until recently, energy development assistance from developed countries was primarily focused on technology transfer and fixed capital assets such as infrastructure, with less consideration on improving self-sufficiency and local capacity building. The majority of development assistance concentrated on pilot projects experiments instead of a genuine interest in diffusing the knowledge or technology to local organizations or communities. The reasons for the concentration on technology transfer could stem from the donor's self-interest to sell their own technology or build markets for export rather than a true concern for the poor (Martinot 2001). Additionally, another criticism is that a sole focus on expanding electrical supply does not solve the issues of affordability or increase the choices of energy services for poor and rural populations (Gaye 2007).

Injustices mentioned previously, which include recognitional, distributional, and procedural injustices all seem to encompass problems such as unequal power distributions and unfair trade-offs of the political ecology framework. A failure to recognize communities' LTK by not adequately consulting with local communities directly impacted by energy development projects has caused problems in other developing countries such as Mexico. One researcher found that limited information resulted in a "lack of engagement with those developing, producing, and selling the technology" so that the consequence was that perceptions grew that the technology was flawed, and so, the public become negatively informed about the RE technology (Mallett 2007). Mallett (2007) argues that the incorporation of LTK can lead to more empowerment of the local people to address local problems and solve them using their own knowledge and skills which strengthens democracy as the voices and concerns of the people are heard. Also, LTK was effective in that community planning resulted in low cost solutions that people followed and took part in. Lastly, justice can also be promoted by the inclusion of the community in order to better redistribute the benefits and costs of renewable energy development. The last impact, justice is especially relevant for renewable energy as the local people could gain control over the energy resources located on their territory and can profit from its use (Corburn 2003).

Recently, community-based renewable energy or community-owned renewable energy are types of adaptive governance approaches where local communities produce their own energy through co-operative structures and gain the benefits from its use. Numerous community-based renewable energy projects have been successful in many parts of the developing world. Cooperative approaches, whereby local civil society organizations partner with governmental bodies as well as NGOs to install RE technologies have been shown to increase engagement and build capacity of local people. This is already being accomplished in several communities in Africa (Munro et al. 2016; Ahlborg 2015). Community solar and other renewable projects produce energy that is locally owned, so this could bypass corruption and corporate profits, as well as having the benefit of communal use. This energy independence could be a way to empower the people, increasing their energy literacy as well as their energy citizenship.

Co-management of resources between government and locals is also promoted by Taylor and de Loe (2012) in their study where they argued that groundwater supplies were improved when the LTK of the people living in a dry region of South Australia was accepted. They argue that collaboration is important because it "raises legitimacy and credibility, lowers disputes, is context specific, links knowledge to action, results in joint fact finding, and is a tool for empowerment (Taylor and Loe 2012)." Another rationale behind adaptive management is that the people on the land have the most knowledge about their resources and have the most to lose if these resources are not managed sustainably. In multiple cases integrating scientific and local knowledge together in resource planning resulted in positive outcomes where knowledge was improved in a general sense and innovation into ways to protect valuable resources was generated (Clark and Brake 2009; Giordano et al. 2010; Kroon et al. 2009). Accruing the support and participation of people on the land is fundamental in order for day-to-day use of resources to be extracted, used, and managed in a way that is sustainable that benefits the people and protects the resource for future use. Managing and sharing the wealth of resources with indigenous communities also aid in the economic development of these communities, so they have the resources to solve their own developmental problems.

Each of the three systems theories offer different rationales for which factors are the most influential to promote sustainable energy development. For socio-technical systems the improvement of electricity access rates is possible with market factors of demand and supply. According to this theory, raising incentives to transition to cleaner technologies by decreasing prices and risks will better permit households to afford to connect to the electrical grid or use propane gas. Furthermore, creating a competitive market to enable new small businesses to enter and promote innovation and investment would improve the ability of the energy market to supply affordable and accessible renewable energy to the population. Finally, promoting energy literacy and skill training of renewable energy technologies and other cleaner and safer cooking energy devices such as improved cook stoves could decrease detrimental health impacts due to IAP. Socioecological systems theories argue that sustainable energy development should encompass effects to the environment and ecological services that life depends on such as air, water, and soil quality that have consequent repercussions of human health and livelihoods. Local knowledge systems such as LTK should be integrated into the implementation of sustainable energy development in order to improve knowledge sharing between professionals and the communities that are more capable of witnessing ecological changes on a local level according to this theory. In order to adapt to changes and adopt sustainable livelihoods local communities need to protect agricultural production and other traditional activities such as fishing. However, as illustrated by the case studies on hydro development which has negatively affected fish stocks, and biofuel production activities such as illegal logging that further exacerbate environmental phenomena such as deforestation, desertification, and salinization improvement should occur to better recognize and protect ecological systems.

Political ecology frameworks demonstrate that injustices are occurring as a result of a renewable energy development in Cameroon. Recognitional injustices seem to be occurring as a result of land grabbing from agrobusinesses and hydro dam development has resulted in flooding that has negatively impacted communities' homes and places of cultural importance such as grave sites. Distributional injustices between the actors that benefit and which actors endure a disproportionate share of the costs of renewable energy development seem to also be occurring in Cameroon with local communities receiving little economic gains from renewable energy development. Additionally, there is unequal distribution of electricity between the rural periphery and urban centers and local communities that are in close proximity to renewable energy projects often lack access to electricity that large foreign-owned companies enjoy such as Alucam. Finally, procedural injustices such as a lack of community consent and failure to implement EIAs seems to have resulted in mistrust and protest towards renewable energy development, especially in the case of Herakles Farm in Cameroon. Therefore, all three theories; socio-technical, socioecological, and political ecology explain some aspect for why the problems with renewable energy development in Cameroon is occurring. Socio-technical market factors can explain the low access rates to electricity due to insufficient electricity supply and inability of consumers to afford the transition to electricity on a household level. Environmental problems and the consequent repercussions to food and water security and its impacts on human health can be explained by insufficient attention to socio-ecological systems in Cameroon. Lastly, political ecology theories convey the problems affecting energy injustices in Cameroon due to power differentials between powerful corporations and elites and small local communities. In order to comprehensively understand the range of factors, actors, as well as the relationships between them inhibit sustainable energy development in Cameroon, all three theories inform the discourse on renewable energy development and each offers solutions to better promote sustainable energy development in the future.

Chapter 12
Recommendations for Sustainable Energy Development

The evidence provided points to the conclusion that the consequences of renewable energy development in Cameroon are caused by integrated factors that affect social, economic, technical, and ecological systems, a concerted approach is therefore recommended to address the problems on a local, regional, and national level. Community, public, private, and international actors and organizations must work together in order to improve renewable energy development in Cameroon to result in a more sustainable and resilient energy sector. Informing decision makers on more effective policies to improve renewable energy development will allow a better mitigation of the negative impacts caused by renewable energy projects in developing countries like Cameroon while increasing access to modern, reliable, and affordable energy services. With this in mind, sustainable energy development is not a fixed destination, but a dynamic evolving process that requires continued learning, adaptation, and cooperation in order to accommodate subjective perspectives on what constitutes as sustainable energy development.

The following recommendations are proposed to mitigate the problems with renewable energy development in Cameroon:

1. Increase the affordability of electricity and gas in order to improve the choice and availability of alternative cooking technologies as well as other energy services such as lighting, cooling, heating, and the ability to power machinery and information technology devices that create opportunities for the creation of entrepreneurial activities.
2. Provide microloans to communities or small-medium enterprises to invest in new technology such as solar panels which can also be used for the dissemination of solar ovens and improved cooking stoves to reduce the consequences of IAP and deforestation on human health and the environment.
3. Develop a solar energy market to take advantage of the vast solar energy potential in Cameroon by providing a favorable business environment through subsides for investment and new entrants in the market, limiting the opportunities of monopoly creation in order to preserve competitiveness, lower prices,

© The Author(s), under exclusive license to Springer Nature Singapore Pte Ltd. 2025
R. Perrault et al., *Sub-Saharan African Sustainable Development*, City Development: Issues and Best Practices, https://doi.org/10.1007/978-981-95-1458-8_12

and result in more decentralized energy systems that are more accessible to rural areas and increases national energy security.

4. Improve community consultation and consent possesses to increase information sharing between local actors and decision makers, and mitigate the negative ecological impacts of renewable energy development which often compound into consequences for human health and local livelihoods resulting a net loss of well-being.
5. Fairly compensate communities that are negatively impacted by hydro and biofuel energy development. Co-develop and finance alternative livelihoods strategies and increase capacity building by training and education investments to improve local resiliency to ecological change.
6. Regulate and enforce laws on illegal logging and other detrimental renewable energy development practices such as agricultural pollutants from biodiesel plantations to better preserve forests and river, lake, and groundwater aquifers in order to reduce food and water insecurity.
7. Raise levels of awareness and education of renewable energy development by offering training on solar energy technology maintenance and aid in the dissemination of mini-hydro stations for community use and management.
8. Cooperate with international organizations, improve governmental integration, and create partnerships with NGOs to increase opportunities for innovation and synergies across private and public sectors.
9. Enhance opportunities for co-management of resources such as forests and adaptive governance frameworks to improve decision making and reduce injustices resulting from inequitable distributions of the benefits and costs of renewable energy development.
10. Invest in opportunities for women and other disfranchised factions of the population in the periphery to improve power differentials.
11. Create new rural energy infrastructure and improve electrical access to social amenities such as health, education, and centers that focus on women empowerment.
12. Foster relationships and encourage learning to broaden participation in renewable energy development in order to promote energy literacy and energy citizenship among the population of Cameroon.

Part III
Urban Development and Sustainability in Nigeria

Chapter 13
Introduction to Urban Development in Nigeria

Nigeria named one of the "MINT" nations (Mexico, Indonesia, Nigeria, and Turkey), is a twenty-first-century African megalopolis inhabited by approximately 200 million individuals, with a population density of 5,926 persons per square kilometre (Akinsete et al. 2014). It is predicted to be one of the future economic powerhouses by 2050 (Akinsete et al. 2014). However, this anticipation of being one of the economic powerhouses has contributed to the nation's rapid growth of unplanned cities which has created numerous problems in Nigeria. One of the most visible and disturbing characteristics of the urban areas in Nigeria is the decline of infrastructural quality. This point is elucidated further by saying that as urban populations grows, and as available resources decline, public infrastructure is being degraded to the point where cities are losing their capacity to operate as productive entities (Innocent 2013). Lanrewaju (2012) noted that the urbanization process in many developing countries particularly Nigeria, has not been accompanied with a corresponding supply of adequate houses and basic amenities. This created a demand for housing, leading to high rents, overcrowding and the development of slums settlements, which also have serious impacts on the built environment and serious consequences on health of city residents (Lanrewaju 2012).

According to World Population Review (WPR) (2017), Nigeria is one of the most populous countries in Africa. Its biggest city, Lagos, has 9 million residents, and a total of 21 million metro area residents (Elisha Sawe 2017). There are also 79 cities with a minimum population of 100,000, and 249 cities with populations that surpass 10,000. This current estimate suggests that by 2050 about 75.42 percent of the country population will be urban dwellers and that some cities, including Aba, Benin City, Ibadan, Ilorin, Jos, Kaduna, Kano, Onitsha, Maiduguri, Ogbomosho, Port Harcourt and Zaria among others will have populations over one million (Aduwo et al. 2016). Table 13.1 below shows Nigerian major cities' populationas of 2017.

Nigeria, with its large diversity in cultural heritage and economic endowments has undergone rapid urbanization with inherent depletion of rural areas with most

© The Author(s), under exclusive license to Springer Nature Singapore Pte Ltd. 2025
R. Perrault et al., *Sub-Saharan African Sustainable Development*, City Development: Issues and Best Practices, https://doi.org/10.1007/978-981-95-1458-8_13

Table 13.1 Total Population of Major Cities in Nigeria 2017

Name	City Population	Increase differences %
Lagos	9,000,000	30.0
Kano	3,626,068	12.1
Ibadan	3,565,108	11.9
Kaduna	1,582,102	5.3
Port Harcourt	1,148,665	3.8
Benin City	1,125,058	3.8
Maiduguri	1,112,449	3.7
Zaria	975,153	3.3
Aba	897,560	3.0
Jos	816,824	2.7
Ilorin	814,192	2.7
Oyo	736,072	2.5
Enugu	688,862	2.3
Abeokuta	593,100	2.0
Abuja	590,400	2.0
Sokoto	563,861	1.9
Onitsha	561,066	1.9
Warri	536,023	1.8
Ebute Ikorodu	535,619	1.8
Okene	479,178	1.6

Source: World population review 2017

of its urban centres in great disarray. The majority of urban workers live with inadequate conditions such as poor roads, lack of water supply, poor sanitary condition, and in most cases, without adequate electricity supply (Abbass 2012). A good example of poor urban planning is the Onitsha megapolis. The filth and infrastructural degradation that have taken over Onitsha is as a result of poor urban planning.

The failure of land sub-division and servicing programs to keep pace with swift urban growth has led to widespread illegal and informal developments, growth of squatter settlements or slum areas. For instance, the housing clusters at Okpoko, Fegge, Woliwo, Odoapku[1] in Onitsha's urban area (Innocent 2013). Other notable cities in Nigeria, such as Enugu, Kano and Ibadan, have a considerable number of slums at the city outskirts. Most of these slums lack health facilities, and other essential infrastructures, hence, the people subjected to such living conditions predisposed to adverse negative impacts from unsustainable urban management that culminates into environmental security issues, such as floods, health implication arising from contaminated water and waste management challenges and food insecurities (Akiyode 2013). There is no doubt that Nigeria as a nation is experiencing hasty

[1] These are slum settlements in Onitsha.

urbanization which has brought about various socioeconomic, cultural and environmental problems, particularly, degradation of the physical urban environment which exists in the nature of loss of biodiversity and climate change, desertification, degradation of agricultural land, air and water pollution, environmental decay and overcrowding (Jiboye 2011).

Social issues faced by urban Nigerians include the lack of job opportunities, which give rise to growing incidence of insecurity resulting from crimes, theft, armed robbery etc., as the only attractive options to those who cannot legitimately sustain themselves (Abbass 2012). According to Gbadegesin et al. (2011), rapid population growth in urban cities like Lagos, Ibadan, Onitsha, Kano, Kaduna, Awka and Port Harcourt has placed immense pressure on solid waste management systems. Solid waste is generated in almost all parts of urban areas and management becomes complicated in larger cities. In most cities, the refuse is dumped in haphazard manner without sanitary landfill. The failure of such systems to cope has contributed to water pollution, environmental degradation, and the spread of communicable disease (Gbadegesin et al. 2011).

Innocent (2003) points out those solid wastes are uncollected and piles of decaying wastes are allowed to rot in the streets; urban roads have deteriorated into quagmires in rainy seasons; public transport systems are becoming severely over loaded; and more are obliged to live in unserviced plot. Not only is little new infrastructures constructed, but existing infrastructures are poorly maintained (Innocent 2013)

Ejaro and Abubakar (2013) notes that the urbanization process in developed countries took many decades, permitting a gradual emergence of economic, social and political systems to tackle the problems of transformations. However, the pattern of urban agglomerations in Nigeria is occurring more quickly against a background of higher population growth and less developed economic, social and political systems. This process of urbanization and city growth in developing countries has been described as "false urbanization". False urbanization means that the process is primarily driven by demographic forces, particularly rural–urban migration, rather than by dynamic economic and industrial forces (Ejaro and Abubakar 2013).

As Babanyara et al. (2010) explains that mankind's activities on the environment in his quest for development has resulted in a continuous and serious degradation of the ecosystem, thus posing a threat to both his present and future living. Human beings use the environment in three basic ways: as a ***resource bank***—the environment supplies them with raw materials needed to maintain their existence, their social and technological structures; as a ***habitat***—people require more space per individual than any other species; and as a ***sink*****for** ***wastes***—human beings produce more waste than other species (Babanyara et al. 2010). This shows that the emergence of urban cities has a great influence on the environment. It further specifies that rapid urbanisation and the growth of megacities, especially in developing countries, has led to the emergence of highly vulnerable urban communities, particularly through informal settlements and inadequate land management (Garschagen and Romero-Lankao 2015).

Most cities in emerging economies are in a poor condition as a result of excessive pressure on existing infrastructures such as housing, roadways, water and energy.

The influx of people into Nigeria cities has drastically impacted the capacity of infrastructures and other several developmental strategies (Jack et al. 2017).

In Africa, South America, and Asia, rapid urbanization has come with complex management challenges in housing, transportation, land use, environmental degradation, crime, unemployment, and public service delivery. These challenges appear to be overwhelming and very difficult to eradicate or control (Aduwo et al. 2016). The issue of unplanned urban development, especially in a country like Nigeria, is alarming because of the concentration of population in urban centres, and the unplanned nature of cities construction by the Nigerian government. Nigeria has undergone unplanned urban development starting from the colonial era[2] in order to meet its counterpart leading to congestion and unhealthy environment. This congestion has led to the massive deterioration of the environment, social problems and unhealthy lifestyles. As Paun (2010) said, hasty urbanization creates significant stresses and strains in many cities.

In the era of globalisation, people are drawn away from their homes to bigger cities due to market availabilities, and urban lifestyle, yet many are disappointed as this massive movement puts pressure on the resources present in those cities. Rapid urbanisation, with its consequent population explosion, has led to an increase in the number of slums in Nigeria cities, for example Lagos.[3]

Considerable changes in environmental conditions have created significant problems for human health, especially vector borne diseases, such as malaria (Oyewole and Awolola 2006). This rapid increase in the population of the urban centres has resulted in an increase in the cost of living. A greater demand on urban limited real estate results in cost of urban land, and high cost of housing, which is often out of the economic reach of the majority of the urban households who incidentally fall into the low-income category. The greater percentage of the poor in the urban area lives in the slum area of the city. This is because substandard accommodations are cheap and such neighbourhoods are in close proximity to their work places (Lanrewaju 2012). The growth of urban poverty has led to the growth of urban slum where most of the components of environmental security are hampered. Three quarters of urban population in most developing economy countries urban centers live in slums. For example Lagos, the biggest city in Nigeria, has more than 100 slums (Akiyode 2013).

Isma'il et al. (2015), notes that housing is a residential structure where man lives and grows. It is therefore universally acknowledged as one of the most basic human needs for survival. The demand for housing has been an issue of global concern as the housing provision still remains one of the most difficult problems facing humanity. It is reported that more than one billion people live without shelter or in unhealthy or unacceptable housing conditions. The most adversely affected lack of housing are

---

[2] Colonial era in Nigeria marked the period of 1914 when the northern and southern part of the country was combined to become a colony under the British government.

[3] Lagos is a southern state in Nigeria with a total population of 9,000,000 and metro population of 21,000,000. It was once the formal federal capital territory of the country and serves as a host of attraction to migrants due to its infrastructural development and presences of sea ports.

the urban poor, who constitute the majority in the developing countries (Isma'il et al. 2015). The misuse use of the natural environment has led to the degeneration of the environment. Such (degeneration) occur as Nigerians attempt to adjust their needs and desires for food, shelter, recreation and infrastructure. These land use activities contribute to the overall development of the country, but they produce a negative impact on the environment, leading to improper resource management (Lanrewaju 2012). One natural resource that is not properly used is sand as it is believed to be in large abundance. It is used daily for construction, and man-made products such as glass etc. This natural resource is used to build urban cities most especially in developing countries. Surprisingly this natural resource is used significantly more than any other except water and air. Sand is the thing modern cities are made of. Essentially every apartment block, office tower and shopping mall from Beijing to Lagos, Nigeria, is made at least partly with concrete, which is composed of sand and gravel together with cement (Beiser 2016).

Another study shows that sand mining is also used for land reclamation and used for city construction. For example, Singapore's high- rise architecture construction is built from sand from beaches and oceans (Comaroff 2014). Sand is also pulled from oceans and rivers in Nigeria and used for land reclamation to build secondary cities for urban development, for example is the Eko Atlantic project which is believed to provide better economic opportunities for the population, development and social recreation, but the reality of this project is that it is will be too expensive for the average Nigerian, since most of the Nigerian population lives under $2 dollars per day. The social attraction that Eko Atlantic project poses is extensive and people are willing to leave their homes to these new cities in search of better economic opportunities. This however leads to the growth of urban slums, and environmental degradation.

However, housing is a necessity of life, rapid urbanization has made access to decent housing elusive to many low-income earners in most Nigerian cities (Aduwo et al. 2016) and this has delivered the notion of building secondary cities (Eko Atlantic project). Beiser (2016), states that to build these cities, people are pulling untold amounts of sand out of the ground, stripping riverbeds, floodplains and beaches which often inflicts terrible costs on the environment since desert sand generally doesn't work for construction. Sand mining is dangerous not only to the environment but also to biodiversity and destruction of livelihood. In India, river sand mining is disrupting ecosystems and killing countless fish and birds. In Indonesia, some two dozen small islands are believed to have disappeared since 2005 because of sand mining, In Vietnam, miners have torn up hundreds of acres of forest to get at the sandy soil underneath and sand miners have damaged coral reefs in Kenya, and undermined bridges in Liberia and Nigeria (Beiser 2016).

A crucial aspect of urbanisation is that city growth and expansion in Nigeria has been largely uncontrolled and thus compounding problems in Nigerian (Opoko et al. 2014). According to UN-Habitat (2017), large urban configurations come with a number of well identified, specific risks: poor urban/regional planning, lack of coordination and deficient coping strategies in the face of social and fiscal disparities (UN-Habitat 2017).

Aduwo et al. (2016), explain that one of the factors responsible for this development is the high rate of rural–urban migration, which has been heightened by the sharp disparities between urban and rural areas in terms of access to job opportunities, basic amenities and services. The rural areas have become increasingly marginalized in national development programmes; many young people in Nigeria have abandoned agricultural production in search of white-collar jobs in non-agricultural sectors creating inadequate job availabilities in urban areas, which has left many migrants unemployed. This has also created more problems, like contributing to a large proportion of low-income earners in urban slums who are unable to gain access to decent housing at an affordable price resulting in homelessness (Aduwo et al. 2016). Another key factor is that it is not just the urban population who suffers, but the rural communities are also are affected.

An analysis of the Netherlands and Japan by Seitzinger et al. (2012), shows that global distribution of land used by the Netherlands to supply food and fibre for the Dutch population require an area four times larger than this small and highly urbanized country. This emphasizes the dependence on agriculture. This shows that with increasing urbanisation, emigration from rural areas to urban centres may not only erode rural communities but also continue to shift the focus of governments away from rural areas; this can create poor governance of the regions which are critical to the successful delivery of resources and ecosystem services to urban areas (Seitzinger et al. 2012). The failure of government to halt these trends of mass movement of people to over-crowd the urban areas with other associated implications is catastrophic. If there is a need to plan and deal with urban problems, it automatically necessitates planning for the rural areas since both areas are symbiotic and interwoven (Abbass 2012). As Samuel said during the Benue state flood, the state also appears to have been drained of its young, vibrant men, who abandoned it for better economic prospects in Abuja, Lagos and elsewhere, leaving behind mostly the elderly and the barely educated young (Ogundipe 2017). The effects of migration to urban cities not only erode government activities from the rural areas, it also devoid the rural areas of human and natural resources needed for economic development and success.

False urbanization is the experience in Nigeria according to Ejaro and Abubakar (2013). There is hardly a city in Nigeria that promotes healthy lifestyles, the urban centres, rather than providing comfort and being centres of arts and civilization, have become urban jungles (Agbola 2005). The challenges to this false urbanization include environmental degradation, deficiencies in urban services, poor sanitation, inaccessibility to key resources, crime and violence (Ejaro and Abubakar 2013). Nigeria's continuous increase in urbanization is related to the lack of access to energy. More than 90 million people, (55%) of Nigeria's population in 2013 have no access to grid-connected electricity. However, the population that can afford backup generators have invested in them in order to have access to electricity which emits carbon dioxide and causes air and noise pollution in cities (Asumadu-Sarkodie and Owusu 2016). High population density and air pollution, especially the severe air pollution many developing cities experience, is posing serious health risks to the urban residents. A recent study in China shows that 83% of the population live in areas where PM2.5 concentrations exceeded the Chinese Ambient Air Quality Standard of 35 μg m −

3, whereas high population-density areas, such as megacities Shanghai, Beijing, and Tianjin, exhibited the highest mortality rates per square kilometer attributed to air pollution (Bai et al. 2017). Poor indoor air quality is associated with respiratory infection, chronic obstructive pulmonary disease, respiratory tract cancers, tuberculosis, cataracts, and asthma. The burning of solid fuels such as dung, wood, and coal exposes people to particulates, nitrogen and sulfur compounds, and benzene. Asthma sensitizers and triggers found indoors include dust mites, cockroach, pet, and rodent allergens, molds, fine dust, tire fragments, and chemical air pollution such as tobacco smoke, wood smoke, and volatile organic compounds from building materials (Ajayi et al. 2014).

These concerns have led to several global summits organized at various levels of government and international agencies including the United Nations. The Millennium Development Goals Summits, the 2002 World Summit in Johannesburg, and the 2005 La-Havana UN sustainable cities documentation of experience programme are such examples. In each of these summits, member nations reiterate the need for good and effective governance as a means of achieving sustainable development in cities (Jiboye 2011).

Chapter 14
Urban Development in Nigeria

Urbanization began during the Industrial Revolution and refers to the increasing number of people who live in urban areas (Chen et al. 2014). Urban centres worldwide are perceived as the centre of excellence, centres of opportunities where aspirations and desires are met (Gbadegesin et al. 2011). Urbanisation brings with it both positive and negative dimensions. Positive elements include internal commerce and foreign trade, financial services and economic growth, growth of modern production and industry, education, and other government services, while negative factors that affect health include increased car ownership which results in traffic congestion and air pollution, lack of green areas, creation of slums (illegal real estate and development), sewage pollution, strain on existing urban infrastructure (e.g. transport, housing and health care) due to a rapid increase in population density, and social problems such as socio-economic inequalities (poverty), prostitution, and crime (Cyril et al. 2013).

According to Fox (2017), early development scholars modelled urbanisation as a response to the growth of jobs in manufacturing and industry, which moved people off farmland into cities, thereby increasing of the total population living in urban areas (Fox 2017). However, further research revealed that people were being ‘pushed’ and ‘pulled’ into urban areas for a wide range of other reasons, such as rural poverty, population pressure, conflict, escape from discrimination or repression, to diversify household income streams, to meet potential husbands or wives, and even simply to find excitement and adventure (Fox 2017). Researchers have also observed that urban development has been historically influenced by both geography and institutions. Living in cities permits individuals and families to take advantage of the opportunities of proximity, diversity, and marketplace competition. For instance, the presence of a port in a city is a major facility able to influence the development process for example through trade relations (Kalu et al.2014).

Cities are generally regarded all over the world as providing the engines of economic development both for the cities themselves and their surrounding rural hinterland (Olujimi 2009). This is the major reason why cities are regarded as areas of

© The Author(s), under exclusive license to Springer Nature Singapore Pte Ltd. 2025

R. Perrault et al., *Sub-Saharan African Sustainable Development*, City Development: Issues and Best Practices, https://doi.org/10.1007/978-981-95-1458-8_14

opportunities for the urban population to exploit for their economic and social development, as well as magnetic areas that attract rural dwellers from their surrounding rural areas (Olujimi 2009). Thus, as the conditions of opportunities expand in urban areas, they worsen in the rural areas as people are continuously attracted into the urban areas where they are dangerously forced to abandon rural life to seek a means of insecure livelihood in towns (Abbass 2012).

In the environment of globalisation, urbanisation has been influenced not only by domestic factors, but also by international factors. Therefore, urbanisation under globalisation has not been an isolated process, but the process and result of a combined effect on multidimensional society, complex economy and integrated culture (Hu and Chen 2015). Paun (2010) puts it that cities are drivers of economic growth (the urban environment is a factor that is supposed to support that function), and that the quality of urban space is an instrument that supports the competitiveness of cities (Paun 2010). Paun analyses cities as physical urban structures, and as public organizations with political decision-making processes which are at the forefront of a range of global governance challenges. Opoko et al. (2014) also explains that cities offer the lure of better employment, education, healthcare, and culture, and they contribute disproportionately to national economies. Under normal circumstances urban development enhances productivity, opens up more economic prospects, creates more wealth, and leads to more innovation that reshapes arts, science, politics, and other professions, but conversely accelerates the spread of diseases and aggravates other vices of exclusion, crime, poverty and environment degradation (Ali et al. 2016).

Main theoretical perspectives by Adams et al. (2016) used to explain the urbanization process is the urban transition theory. The urban transition theory builds on research claiming an association between urban environmental burdens and growing affluence. Urban areas are an underutilized resource that concentrates much of the countries' physical, financial, and intellectual capital, in the process of wealth accumulation and environmental challenges, become more dispersed, delayed, and shift in type, thus as cities become more urbanized and industrial activity increases, environmental degradation could occur because of the increased energy use and emissions. However, this negative effect could be reduced by the implementation of energy efficient environmental regulations and technological innovations by government policies if well implemented (Adams et al. 2016).

As Nwokoro and Dekolo (2012) explains, cities expansion in developing countries and the global south is characterized by sprawling fractals or spider-like configurations rather than compact aggregations seen in the developed world. The consequence of the sporadic growth and sprawling development especially in rapidly growing metropolitan regions is reflected in problems like climate change, resource depletion, and diminution of agro-ecological and rural lands (Nwokoro and Dekolo 2012). The urbanization process in developing countries is occurring more dramatically and rapidly compared to that in developed countries (Chen et al. 2014). For developing countries, these urban transitional periods pose numerous urban social challenges for many of its dwellers, with many of these cities transforming into informal urban expansion, urban disintegration and segregation with the neglect of the larger population becoming a common phenomenon (Adeola 2016).

This urban phenomenon is continuously increasing, extending the boundaries of the city or metropolitan areas. Thus, urban planning is experiencing a crisis; at least the concept of urban planning that appeared with the modern movement and its myriad of architects. The reality in the big cities in Nigeria such as Lagos, Ibadan, Port Harcourt and Benin presents a number of problems that are worth mentioning. These include urban decay, overcrowding, lawlessness, loss of land and natural resources (Gbadegesin et al. 2011). In Nigeria, the growth and complexity of human settlements and particularly the process of urbanisation have been prodigious since it is increasing at a greater rate than the capacity of urban planners; this poses significant quality of life challenges such as rising inequality (Ekong 2017). Today, Nigeria has some of the world's most polluted cities and most steeply rising greenhouse gas emissions, for example in Lagos, Aba and Port Harcourt.[1] The implications of the rapid urban growth have been described as varied with diverse consequences that are clearly seen in inadequate housing and infrastructure; an economy dominated by the informal sector activities, rising crime rates and inadequate water supply, waste disposal, traffic and human congestion (Ekong 2017).

Some scholars have long recognized that if the less developed nations follow the development trajectories of the more developed nations it will be unsustainable on a planetary scale (Givens 2015). African Urbanisation is described as pseudo- urbanisation by Opoko et al. (2014) in which many developing countries have followed a different trajectory from the above premise, leaving many overwhelmed urban residents and their governments in frustration, despair and confusion (Opoko et al. 2014). However, this rapid growth in urban population accentuates the demand for natural resources, leads to changes in land use patterns, causes pollution and loss of biodiversity, and alters hydro-geomorphology (Rajashekariah 2011). Urbanisation impacts wetlands in numerous direct and indirect ways. For example, construction reportedly impacts wetlands by causing direct habitat loss, suspended solid additions, hydrologic changes, and altered water quality indirect impacts, including changes in hydrology, eutrophication, and sedimentation which can alter wetlands more than direct impacts, such as drainage and filling (Ajibola et al. 2012).

Rajashekariah predicts that most of the urban growth is expected to occur in regions around the world's biodiversity hotspots, and human settlements represent the most profound alterations to the natural environment (Rajashekariah 2011). Land use also has other dire consequences on other natural resources including water. This poses a serious threat to the population living in urban centres, especially those in slums. Another resource used greatly to meet urban needs is sand. As land reclamation activities continue in a widespread manner, the natural resilience of the land to flooding is reduced, as highlighted by Idowu et al. (2017). Thus, wetlands livelihoods may be under threat in case of extreme rainfall (above 50 mm in a single downpour). Furthermore, natural processes and sediment deposition are hindered as reclaimed lands are cut off from an estuary or the sea, thereby hindering natural land tides (Idowu

[1] These cities are in the southern part of country.

and Home 2017). A study shows that in Nigeria land reclamation and sand filling of waterfronts, swamps and the Lagos lagoon for residential development is a practice by the Lagos State government and individuals that has been widely condemned. Some high profile estates in Lagos were sand –filled, and the government is presently proposing reclaiming part of the Atlantic Ocean for its mega project, "Eko Atlantic City" (Nwokoro and Dekolo 2012). In other words, chances of wetland submergence against the increasing sea levels become high due to the extensive land reclamation activities and the indiscriminate dredging of sand for land reclamation in Nigeria. This has a tendency of disrupt to hydraulics of the lagoon, and increase flood risk in certain parts of the wetlands (Idowu and Home 2017).

In a rapidly urbanizing country like Nigeria where about 50% of its 140 million people live in urban areas, increasing population without corresponding economic growth, physical development challenges, global climate change as well as inability of institutions to integrate disaster preparedness, prevention and mitigation strategies into its national development plans and programs, have been identified as responsible for increasing vulnerabilities in urban areas in this country (Ibem 2011). Increased urbanization is a major concern for less developed countries since they often lack the infrastructure and basic services (e.g., water, sanitation and healthcare) necessary to absorb the increasing number of people (Mahabir et al. 2016). As Abbass (2012) explained, such rapid urbanization in Nigeria has its inherent negative implications in social, environmental, economic and political dimensions, While the social implication of congestion and over-crowding are obvious or certain the inadequacy of all sorts of services cannot be ruled out. The conglomeration of people from diverse ethnic and cultural groups also brings about struggles and conflict between them in the political or even economic arena, which raises tensions between the interested parties (Abbass 2012). Societies constantly experience changes, and they inevitably affect the physical development of cities and towns. If developments in these centres are not properly handled accordingly, this could lead to degradation of the physical environment, and destruction of the social elements (Gabriel and Abraham 2011).

In many developing countries, rapid urbanisation has placed severe pressure on urban housing, pushing costs up and, thus, constraining affordability. This urban challenge has increasingly complicated, worsen environmental and human settlement problems in Nigeria (Nwuba et al. 2015). Affordability has been one of the issues at the core of urban housing problems (Nwuba et al. 2015). Unplanned urban pressure and demographic growth has promoted deforestation, air pollution, flooding and have altered the pattern of the existing ecosystem. The housing has been universally acknowledged as one of the most essential necessities of human life and is a major economic asset. Adequate housing provides the foundation for stable communities and social inclusion (Amao and Ilesanmi 2013). Consequently, Nigeria has one of the highest urban growth rates, having cities ranked among the fastest growing in the world, however, rather than improving the physical condition of the urban space, cities continues to degenerate while investment in urban infrastructure including housing, has failed to keep pace with the growing population.

A major reason attributed to these urban developmental problems in Nigeria is the lack of effective mechanisms of governance needed to ensure a rational level of equilibrium between the growing population, the cities environment and existing infrastructure, hence, the increasing poor quality of the living conditions and low liveability index of urban areas in Nigeria (Jiboye 2011). As Amao and Ilesanmi (2013) explain, the housing situation in Nigeria is characterized by some inadequacies, which are qualitative and quantitative in nature.While the quantitative housing problem could be solved by increasing the number of existing housing, the qualitative inadequacies are enormous and complex. Despite Federal Government access to factors of housing production, the country could at best expect 4.2% of the annual requirement. Market failure to provide affordable housing has created problems for households living below the poverty line by forcing them to occupy low-quality and overcrowded dwellings located either in decaying areas within the central city, or in informal settlements located at the urban fringe. Inadequate housing affects a large proportion, of all urban residents in the developing world. Moreover, poor urban infrastructure and deficient services, the absence of formal land titles, and generally poor local environmental conditions are now common characteristics of cities in developing countries (Amao and Ilesanmi 2013). Therefore, it is important to look at the progression of urban development in Nigeria.

Chapter 15
Process of Urbanisation in Nigeria

Although urbanization and urban cities existed in Nigeria before the colonial period, it was at a small scale such as centres of barter trade (Kalu et al. 2014). It is important to note that Nigeria was predominately rural even before the arrival of the Europeans. Before the advent of colonialism, which gained full-fledged prominence in the late nineteenth century, the country had some flourishing cities, kingdoms, empires and towns that can be likened to the modern-day urban centres which included the Sokoto Caliphate,[1] (Fig. 15.1) the Old Oyo Empire,[2] the Benin Kingdom[3] (Fig. 15.2) and the Nupe Kingdom,[4] to mention but a few. Conversely, however, these towns and cities are somewhat different from the modern-day megacities in that they did not bring about huge social inequality (Akanle and Adejare 2017). The pre-colonial towns were self-sustaining in all respects: sufficient in food supplies with large track farming areas within the cities; large markets; effective local processing, transporting, storage, wholesale and retail purchasing (Abbass 2012) which gave rise to the town development. The northern savanna cities grew within city walls, at the centre of which were main markets, government buildings and the central mosque (U.S. Library of Congress 1991). This system was oriented around trans-Saharan trade. The Hausa States were part of a trade network stretching across the Sudan region northwards

---

[1] The Sokoto Caliphate was an independent Islamic Caliphate, in West Africa and founded during the jihad of the Fulani War in 1809 by Usman dan Fodio, it was abolished when the British defeated the caliphate in 1903 and put the area under Northern Nigeria (Harvard Divinty School 2018).

[2] Oyo Empire (1600–1836) was a Yoruba empire, located in southern part of the present Nigeria. Oyo rose to become the most important city state, surpassing Ife during the 1300s and is presently located in the northern part of Lagos (Afropedea 2010).

[3] Benin Empire was located in southern Nigeria with Edo (now known as Benin City). The Benin Empire was "one of the oldest and most highly developed states in the coastal hinterland of West Africa, dating back to the eleventh century"(Ogunsi 2016).

[4] The Nupe Kingdom was one of the most prominent Kingdoms of the Niger-Benue confluence area, what is today known as the Middle Belt of Nigeria. At its peak, the Kingdom spanned several Nigerian states especially Niger and Kwara states but its influence was felt through the entire Middle Belt (Eneke the Bird 2014). It is located in the northern part of Nigeria.

© The Author(s), under exclusive license to Springer Nature Singapore Pte Ltd. 2025

R. Perrault et al., *Sub-Saharan African Sustainable Development*, City Development: Issues and Best Practices, https://doi.org/10.1007/978-981-95-1458-8_15

Fig. 15.1 Sokoto Caliphate 19th Century. *Source*: Sokoto Caliphate, Nineteenth Century

Fig. 15.2 Benin Empire. *Source*: Omar Alleyne Lawler, 2015

to the ports of North Africa and Europe (Bloch et al. 2015). The concentration of wealth, prestige, political power and religious learning in these cities attracted large numbers of migrants, both from the neighbouring countryside and distant regions (U.S. Library of Congress 1991). Islam was for quite some time the religion of the court and commerce, and was spread peacefully by Muslim clerics and traders, and which led to the growth of old northern cities like the Sokoto Caliphate, Katsina and Kano[5] (Huud 2015).

As part of the administrative organisation after the conquest of the Sokoto caliphate, the British created new administrative headquarters that led to the establishment of new urban centres like Kaduna, Sokoto, Maiduguri and Yola, and the expansion of existing ones like Katsina, Kano and Zazzau (Shehu 2017). The colonial government also adopted the Garden City approach in the planting of European Quarters, later Government Reserved Areas (GRAs), with spacious, lush greenery especially in places like Lagos (Ajayi et al. 2014). Colonial rule contributed to the development of a number of urban settlements, especially along the evolving rail and road-river networks in eastern Nigeria. Such centres include Port Harcourt, Aba,

[5] These old cities are located in the northern part of Nigeria.

Enugu, Owerri, Umuahia, Okigwe, Calabar and Onitsha.[6] By virtue of their accessibility, each of these serves as a point of trade articulation for their tributary areas in the expanding international economic relations involving the export of agricultural produce and minerals, and the import of manufactured goods. Such roles have immense population pulling potential. With such a head start, these centres have continued to attract the bulk of the modern urban infrastructure, and of the industrial and commercial activities (Kalu et al. 2014).

However, pre-independence Nigeria had a very low urban population as well as urban population growth in 1921, only ten Nigerian settlements had urban status according to Opoko et al. (2014) and by 1931 only two cities, Lagos and Ibadan had populations in excess of a hundred thousand (Opoko et al. 2014). In 1952, 10% of the population lived in urban centres with a population of 20,000 people and above which increased to 20% and 38% in 1970 and 1993 respectively. By the year 2010, an estimated 60% of the population lived in cities. The growth in the size of cities has been equally rapid, in 1960; Lagos and Ibadan were the only two cities with more than 500,000 people (Lanrewaju 2012).

Another factor was the attainment of Nigeria independence in 1960 and the first creation of states in 1967[7] which led to increased migration for economic gain and job opportunities (Shehu 2017). Aduwo (2016) also noted that government's deliberate policy on decentralization of administrative structure through the creation of States and Local Government Areas, and the establishment of four main industrial-urban conurbations: Lagos- Ibadan, Kano-Kaduna-Zaria-Jos, the Benin-Sapele-Warri and Port Harcourt-Aba-Onitsha- Enugu are major factors (Aduwo et al. 2016). The massive improvement in road development and availability of vehicles also promoted urbanisation (Ekong 2017). Extensive development is a feature of the nineteenth and twentieth centuries, with concentration of economic and administrative decision-making in Lagos, Ibadan, Kaduna, Jos, and Enugu, and larger populations associated with greater specialization of goods and services led to a massive movement of populace from the countryside to the cities (Oyeleye Oyewale 2013).

The oil boom[8] of the 1970s (Jiboye 2011; Abbass 2012; Ekong 2017) also provided funds for major urban infrastructural development across the country and was instrumental increasing the tempo in rural–urban migration. This increased the concentration of the population in urban areas in early cities, like Lagos, Kano and Port Harcourt (Aduwo et al. 2016). The oil sector provided 90% of the government revenue, a noticeable rise in industrial investment both in the public and private sectors. This led to the balance of payment surplus, growth in construction industries, rapid urbanization and ostensible advancement in educational, health and infrastructural development.

[6] Thes towns are the southern part of Nigeria.

[7] 1967 marked the first creation of states in Nigeria by the military head of states Gen. Yakubu Gowon. 12 states were created in this period.

[8] The oil boom of the 1970s shifted the country from its agrarian system to non-traded sector in petroleum thereby proving the country with lots of revenue and economic development. The oil of the 1970 made Nigeria one of the top ten oil exporters in the world.

In fact, by the time the Third National Development Programme[9]was launched in 1973, it was said that money was no longer a problem in the economic development of Nigeria but how to manage it (Ayedun et al. 2011).

During the 1970s, Nigeria was confronted with the challenge of reconstruction, rehabilitation, and reconciliation in war-ravaged areas. However, other socio-economic developments were designed and executed throughout the country. This no doubt involved the state's direct involvement in the resettlement and movement of people from different parts of the country (Abbass 2012). The establishment of industries also contributes to the physical expansion of an area. One of the major industries established was Hamada Carpets in 1978,[10] followed by the Katsina Steel Rolling Mills in 1982[11] (now Dana Steel Company) (Shehu 2017) which also created a rapid population explosion, urbanization and rural depletion. Such mobility trends in the population had adverse consequences for the space economy and for the people themselves both in towns and the country, thereby changing population characteristics in both the rural and urban habitats which had their own distributional dynamics and impacts (Abbass 2012).

The 1963 National Population Census[12] classified the urban centres as being settlements with a population of more than 20, 000 (Akiyode 2013). According to the final figures of the 1991 census, the population of Nigeria however, was 88.92 million. Projection of the population, using a 3.0% growth rate, showed that the population of Nigeria could be about 106 million in 1999. The Population Reference Bureau estimated the total population of the country in 1999 to be about 113.8 million (Lanrewaju 2012). The city of Lagos, for instance, has continued to grow in size since the 1960s with an annual growth rate close to 14% during the 1970s when the massive extent of new construction was exceeded by the influx of migrants attracted by the oil boom (Jiboye 2011).

A study by Opoko et al. (2014) shows that 43.5% of the Nigeria people lived in urban areas in 2000 with predictions that the urban population would hit 65% by 2020, given an annual urban population growth of 5.5% (Opoko et al. 2014). Oyeleye (2013) explained that rural–urban migration in Nigeria has increased over the past 25 years. Between 1985 and 1990, over 3 million Nigerians migrated from rural areas to urban centres, while over 5 million Nigerians migrated between 2001

[9] The Third National Development Plan of the 1975–80 is the third in the series of national development drawn in Nigeria. It is the biggest developmental plan drawn in Nigeria since the 1960 and had a budget of 30 billion Naira (Lewis 1977).

[10] Hamada Carpets is an industry located in Kaduna which is in the northern part of Nigeria. It was founded in1978 and promoted the rural–urban migration to Kaduna state for employment.

[11] Iron and steel industry in Nigeria developed as a public funded integral industry. Between 1979 and 1983, the Nigeria government jump started iron and steel production with emphasis on the importance of iron and steel in developing and driving local production of goods. A strong iron and steel industry was also projected to reduce demand of foreign currency used towards the importation of steel products (Agbu 2007).

[12] Following the cancellation of the 1962 census result, another head – count was ordered in 1963. It took place in November 1963, in a tense political atmosphere occasioned by increased awareness that political representation control and revenue allocation were determined mainly by population.

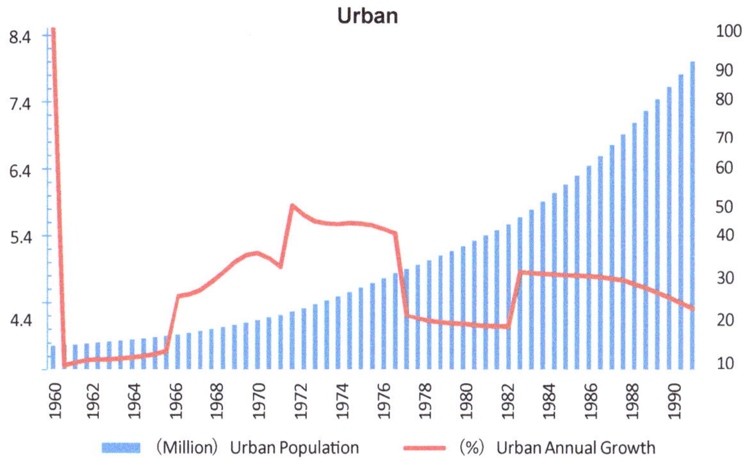

Fig. 15.3 The growth of Urbanisation in Nigeria from 1960 to 2016. *Source*: World Bank Data (2016)

and 2005. This is greater than a 75% increase in the rural urban migration in Nigeria. This rural–urban drift will continue to increase in Nigeria if physical developments continue in urban areas, while rural communities are left unattended in terms of the development of infrastructure, services, social amenities and economic activities (Oyeleye Oyewale 2013).

The above review has shown that three major factors contributed to the urbanisation process in Nigeria: the creation of GRA by colonial governments, the oil boom and independence of the 1960s, and the creation of states. This increased urban development and migration in Nigeria. It is equally important to note that after the colonial government administration, no structural adjustment plan was made for urban development, thereby leading to the major urban issues that Nigeria has today. The graph below describes the growth of urban population in Nigeria from the 1960 to 2016 (Fig. 15.3).

In the above table, it is shown that urban growth in Nigeria has increased from the period of independence in the 1960s from 15 to 40% to 2016 due to many factors such as the oil boom and industrialization.

Chapter 16
Causes of Urbanisation in Nigeria

For many years, rural–urban migration was viewed favourably in economic development as a result of industrialization. Internal migration was thought to be a natural process in which surplus labour was gradually withdrawn from the rural sector to provide needed manpower for the urban industrial growth process. However, today migration has become a major factor contributing to the excess labour force in the urban centres, which has continued to increase the rate of unemployment and other social challenges caused by the imbalances between rural–urban environments (Kalu et al. 2014). The trend in Nigeria's oil economy is characterized by rural exodus, Nigeria's rural oriented economy with the vast majority of the population expanding their labour force in agricultural pursuits is nonetheless characterized with migration economy based on the long history of Nigerian urbanization (Abbass 2012).

According to many studies used to examine the economic effects of rural–urban drift on the migrants, the points of the origins and the destinations reveals that internal migration has a negative impact on the quality of rural life because it reduces the number of individuals in rural areas (Oyeleye Oyewale 2013).

> Asaju gave a vivid picture of the situation in the rural areas in Nigeria thus: There was a feeling of desertion, loneliness and deprivation clearly written on the faces of the rural dwellers.... The villages are thinly populated with old people and children. The rush to urban centre brought about by oil boom of 1970's had attracted young men and women away from home (Asaju 2017).

Many factors contributed to the rural–urban migration in Nigeria. The first factor is colonialism which played a role in rural–urban migration. It exposed Nigerians to the international market by the introduction of cash crops,[1] and creation of administrative offices in planned zones. This necessitated the investment in socio-economic amenities within these areas. This stimulated the quest for paid employment and migration of workforce into these areas and therefore prompted the commencement

[1] Cash crops are crops produce by farmers in Nigeria which are used for export. It was the source of revenue for the Nigerian government before the oil boom of 1970.

© The Author(s), under exclusive license to Springer Nature Singapore Pte Ltd. 2025

R. Perrault et al., *Sub-Saharan African Sustainable Development*, City Development: Issues and Best Practices, https://doi.org/10.1007/978-981-95-1458-8_16

of rural–urban migration, especially into new cities such as Lagos, Port-Harcourt, Enugu, Jos and Kaduna since unskilled labours were required for menial jobs in the urban centres of colonial administration (Okhankhuele and Opafunso 2013). The rural economy is principally agrarian, but, under conditions of increased population and diminished rural resources, resulted in rural population migrating to urban areas for job security, higher education, higher income, better health and a longer life (Lanrewaju 2012). The second factor leading to rural–urban migration is the neglect of the infrastructure of rural areas. Many people may move to the city for better economic or educational opportunities due to a lack of markets, good transportation facilities, schools, and health facilities. From Kalu et al. (2014), study of rural infrastructure in Nigeria, they found out that wide urban–rural disparities were a major reason for the massive rural–urban migration of the 1970s (Kalu et al. 2014). A study by Asaju (2017) also explains that failure of the government to implement strategic policies necessary for rural development in Nigeria was not transparent and feasible enough to carry out its obligations. This study shows that agricultural practice which was the major occupation of the rural communities was not properly enforced by the government. This resulted in the underdevelopment of the rural communities in Nigeria (Asaju 2017).

The third reason people choose to migrate is due to agricultural decline affected by government policy. As Abbass (2012) explained, the structure of Nigeria's rural economy has made it even more vulnerable to increased state intervention in the rural agricultural economy, particularly through the introduction of a variety of agricultural programmes and strategies. Thus, with large tracts of land forcefully acquired by the state and held by the urban bourgeoisie in direct league with the foreign partnership, these lands have been transformed into commercialized and mechanized farming, which makes land acquisition difficult. The rural inhabitants are inevitably transformed and turned into wage labour captivity particularly in the River Basin and Development Authority (RBDA), and Agricultural Development Project (ADP) areas where land had been acquired for diverse purposes (Abbass 2012). The Federal Military Government established ten River Basin Development Authorities. These were followed, in August 1976, by the Niger Delta Basin Development Authorities, bringing the total number of River Basin Development Authorities in Nigeria to eleven (Adegeye 1982).

The RBDAs[2] were established to provide water for irrigation and domestic water supply, improve navigation, hydroelectric power generation, and the recreation of facilities and fishery projects. The basins were also expected to engender big plantation farming and encourage the establishment of industrial complexes that could bring the private and public sectors in joint business partnership and to bridge the gap between the rural and urban centres by taking development to the grassroots and discourage migration from the rural areas to the urban centres (eWASH 2009). The

[2] The RBDA scheme in Nigeria failed to deliver its promise due to multinational companies involvement, poor government policy especially in place of poor local population, Local populations who bear heavy social costs of the developments including loss of homes, fertile farm lands, fishing and fishing grounds environmental problems from dams constructions, lack of funds and revenue.

Agricultural Development Projects (ADPs)[3] were designed in response to a fall in agricultural productivity. They also addressed the concern to sustain domestic food supplies as labour had moved out of agriculture into more remunerative activities, that were benefitting from the oil boom domestic recycling of oil income provided the opportunity for the government, with Bank support, to develop the ADPs with commitment of $1.2 billion for Agricultural renewal project (Independent Evaluation Group (IEG) 2016). With agricultural pursuits becoming more unaffordable by the rural peasantry due to, among others things, high costs of inputs, fertilizers, and labour the peasants have, by and large, been systematically forced out of farming and to abandon farming by selling or mortgaging their farmlands to move out of the rural environment for the frivolous urban life, for wage labour and other lumpen-proletarian works (Abbass 2012).

Some people may decide to migrate because their rural economy is disrupted. Such decisions could be as a result of natural catastrophe such as floods, droughts, landslide erosion earthquakes, insect and pests infestations, escape from lack of human rights and justice, political instability, infertile soil, lack of arable land for cultivation, communal clashes, family dispute, outbreak of war, and other adversities (Okhankhuele and Opafunso 2013). Okhankhuele summed these up by saying:

> The factors influencing the decision to migrate are varied and complex: Emphasis has variously been placed, for example, on; Social factors; including the desire of migrants to break away from traditional constraints of social organizations; Physical factors; including climate and meteorological disasters like floods and droughts; Demographic factors; including the reduction in mortality rate and the concomitant high rates of rural population growth; Cultural factors; including the security of urban 'extended family' relationships and the allurements of the so-called 'bright city lights'; Communication factors; resulting from improved transportation, Urban-orientation; educational systems and the 'modernizing' impact of the introduction of radio, television and the cinema (Okhankhuele & Opafunso 2013).

The fourth reason was also wage differences. Rural–urban wage gaps reflected not only productivity differences but also artificially high wages that attracted many migrants. Rather than offering economic benefits, migration to cities and the eventual closing of the wage gap merely resulted in more workers waiting through unproductive spells of unemployment or underemployment in a bloated service sector (Oyeleye Oyewale 2013). As Jiboye (2011) points out, there is a need to integrate rural development plans into the national strategies in order to bridge the wide gap between the urban and rural areas in terms of job opportunity and availability of basic infrastructure (Jiboye 2011). It is worth noting that characteristics such as age, gender, ethnic background, socio-economic status, educational status, and religion influence the decision to migrate to the city. For instance, a study in Anambra State (Eastern Nigeria) found that many Igbo families encourage members to migrate, believing that staying in the village or rural area will not bring financial success (Kalu et al. 2014).

---

[3] The decline in oil prices that began in 1982 had a substantial fiscal effect in Nigeria and led to shortages of counterpart funds for these projects and farming scheme was more based on high input technology which was not used by smallholders and local people and poor implementation on the part of government.

As discussed above, the factors responsible for rural–urban migration in Nigeria are deemed economic factors which are related to livelihood, social factors due to lack of infrastructural development, historical factors due to the nature of colonialization and environmental factors relating to floods, and ecological constraints. These factors affect each other and have promoted segregation between the urban and rural populations. The economic value placed on agriculture and industrialization as a means of growth have helped to determine the former as a lesser source of development and the latter as a means of growth. A good example is an emphasis placed on white collar jobs and non-white collar jobs which relates to wage difference thereby creating a gap between urban and rural populations. In cities, rural spaces, and the global economy, distributional inequalities, including those of the environment, have a demonstrable spatial expression and constitution (Walker 2009). Although this segregation is not only between the urban and rural paradox, it is present in the cities themselves and is explained through the various theoretical framework discussed in the following chapter.

Chapter 17
Urban Theories Explaining Nigeria's Situation

17.1 The Urban Political Ecology Theory (UPE)

The Urban Political Ecology Theory (UPE) is a branch of political ecology that emerged in the late 1990s rooted in Marxist-structuralism tradition that opens up politicizing views not just on the complex dialectic relationship between society and nature but also about the deep roots of power-relations, neoliberal metabolic processes, uneven socio-ecological development, and patterns of winners and losers embedded in the reproduction of urban spaces and societies (Ajibade 2017). It builds on the political analysis of the socio-ecological relationships and power struggles which underlie environmental transformations (Ajibade 2017). This normative concept and social movement in the United States was used to explain the highly uneven distribution of environmental 'goods' and 'bads' in the city (Swyngedouw and Kaika 2014).

According to Kaika et al. (2014) early work in the 1980s began to recognize that poor, often predominantly African American, neighbourhoods were overwhelmingly located in areas that were characterized by high concentrations of environmentally hazardous conditions, like the presence of toxic dumps, waste processing facilities, ground pollution, hazardous chemicals, and the absence of green zones etc. on the one hand, and concentrations of low-income households on the other. In other words, the spatial distribution of environmental goods (green amenities, good ecosystem) and bads (pollution, flooding, traffic noise) mirrored the socio-spatial distribution of wealth and income, and it became clear that sustainable urban lives are primarily the privilege of the rich (Swyngedouw and Kaika 2014). In essence, many contemporary theories of justice refer to a standpoint that is broader than just how things are distributed, yet for all of these developments in justice theory, very little has been applied to the environmental justice movement. Most discussions of environmental justice focus on maldistribution—the fact that poor communities, Indigenous

© The Author(s), under exclusive license to Springer Nature Singapore Pte Ltd. 2025

R. Perrault et al., *Sub-Saharan African Sustainable Development*, City Development: Issues and Best Practices, https://doi.org/10.1007/978-981-95-1458-8_17

communities, and communities of colour get fewer environmental goods, more environmental bads, and less environmental protection (Schlosberg 2007). Other distributional inequalities may compound these or, crucially, may constitute the basis for environmental injustice claims even where seemingly equal and even socio-spatial patterns of environmental goods or bads are observed (eg an "equal "distribution of pollution, flood risk or green space) (Walker 2009).

Emphasis is put on the ethnically and economically uneven positions in the political and economic decision-making machinery that allocates the distribution of goods and bads throughout the city, revealing that environmental 'goods' are partitioned such that urban elites benefit most while environmental 'bads' are left to the areas of the powerless and disenfranchised thereby creating unequal distribution (Swyngedouw and Kaika 2014). This socio-ecological process changes the social reproduction of urban spaces and the social relations of property in ways that increase resilience for some (powerful individuals) while expanding vulnerability and risks for others (especially the poor and marginalized) (Ajibade 2017).

17.2 Marxist Theory of Capitalism

In Marx' analysis of capitalism, the conception of the class involves a collective group of individuals who share similar economic and social relations relative to each other in a society where relationships can be objective or subjective. Marx went further to argue that the first criterion underclass system is the division of a society into the owners and non-owners of means of production which are capitalist (bourgeoisie) and proletariat that include self- employed artisans, small shopkeepers, and many professionals who do not have ownership of the means of production and whose only means of subsistence is to sell their labour power for a wage or salary. From the Marxist perspective, capitalism is a system based on the exploitation of the proletariat by the bourgeoisie (Omotosho 2015). In a similar Marxian framework, on dependency theory, to whit world system theory (WST), of which Immanuel Wallerstein is especially prominent, using an explanation for the emergence of slums in megacities to explain this theory. According to this theory, the history of the world is that of domination such that the haves dominate the have-nots; the periphery countries are dependent on core countries for capital and have underdeveloped industry. Dependency here connotes weakness on the part of one and strength on the part of the other in a symbiotic relationship. How this relates to the explanation of the emergence of mega-slums is that mega-slums do not exist on their own – they are produced by the complex social situations that characterize megacities (Akanle and Adejare 2017).

Using illustration from Omotosho (2015), the social class structure put forward by Marx is used to explain the urban spaces during the colonial and post-colonial era in Nigeria were a reflection of the class system such that there were distinctions within residential patterns based on relations to the means of production. Elites were known to occupy certain parts of the city that were usually regarded as Government Reserved Areas, estates, villas and so on, which has constant electricity supply, good road and telecommunication network, and adequate security to mention a few, and on the other hand, city centres were usually the domain of the lower-class members; and some of the features include private police patrolling the area around the clock, making water available only at public fountains, no sanitation systems, electricity is pirated by a privileged few, roads become mud streams in the rain, and where house-sharing is the norm. However, this may no longer hold as elites have further extended their spheres of influence within cities to the city centres by dispossessing urban poor from these spaces. Old and indigenous houses are being bought and converted to commercial centres, political office holders are partnering with private investors to perpetuate their interests within city centres, and the poor are further pushed away from these spaces to the suburbs without due recourse to the law (Omotosho 2015). This shows a clear distinction between separation of social inclusion and exclusion. As Amao and Ilesanmi (2013) emphasized, the social and financial distance between the urban poor and the economically better-off classes are reflected in the limited opportunities for the former to obtain housing through anything other than informal means. This limitation has produced a highly uneven and segregated distribution of housing quality across socio-economic classes and also over space. The most negative effect of housing supply shortages and housing inequities in Nigeria is the formation of slums at the fringes of virtually all urban areas. These areas are typically occupied by recent rural migrants and other low-income groups and are often abandoned in terms of political and economic support (Amao and Ilesanmi 2013).

17.3 Neo-Liberalisation Theory

Another theory used to explain the class system is the neo-liberalisation of African cities. This policy has been used to overcome the legacy of urban decay in African cities through the development of major cities example the Eko Atlantic City project. This not only perpetuates more problems for urban development but also creates segreations in cities leading to unhealthy lifestyles. As Carmody and Owusu (2016) explain, mega urban developments and attempts to formalize the informal sector represent neoliberal approaches to the urban crisis in Africa. Eko-Atlantic, currently under construction on reclaimed land off the coast of Lagos, has a similar utopian/dystopian or heterotopia vision. Lagos, a mega-city with an estimated population of over 15 million, is facing many severe urban problems. Its burgeoning population is marked by high rates of poverty and income inequality. The city is also known for urban decay as reflected in overcrowded housing, congested traffic, inadequate sanitation and social services, and a high unemployment rate. Despite these problems, this city remains the economic pulse of the largest economy in Africa. In sum, these planned urban developments can be characterized as utopian dystopias not only because they are unlikely to generate substantial economic benefits in the form of job creation, but also because through their representations, micro-spatialities and external foci, they discursively erase the deepening contradictions of the urban agglomerations with which they are associated. As such, they can be seen as yet other examples of the equalization of space and class associated with global neoliberalism, even as they attempt to reverse marginalization through greater integration into the global economy. The new exclusions that they generate may further deepen the structural violence experienced by the majority of the population (Carmody and Owusu 2016).

Nigeria's reality, however, is a juxtaposition of extremes, with the city of Lagos a living hyperbole of this inequity, while a proportion of the city lives in the lap of luxury with multimillion-dollar properties and yacht clubs, a majority of the population lives far below the poverty line with incomes of less than US$1.25 per day (Akinsete et al. 2014). As Plastow (2010) explained the traditional approach in environmental governance has been government regulation, in what is known as 'command and control' practices such as capping emissions or fish catches, which rely on legally established limits and are enforced by the state (Plastow 2010). However, many policies today contain elements of the neoliberal agenda and what has been called 'free market environmentalism' has proliferated in a dialogue between environmentalism and proponents of neoliberalism, include tradeable emission permits, eco-taxes, transferable fishing quotas, user fees for public goods and aspects of utility privatization, as well as plenty of corporate green-washing (Plastow 2010).

17.4 Theoretical Framework

Figure 17.1 below represents the conceptual model of this study. The model seeks to simplify the theoretical framework present in the literature earlier in this section. It shows how political and economic conditions affect social lifestyles of the urban citizens which in turn negatively affect the environment. The model presents the power struggle between the privatization (economic) and public (government) in the distribution of resources.

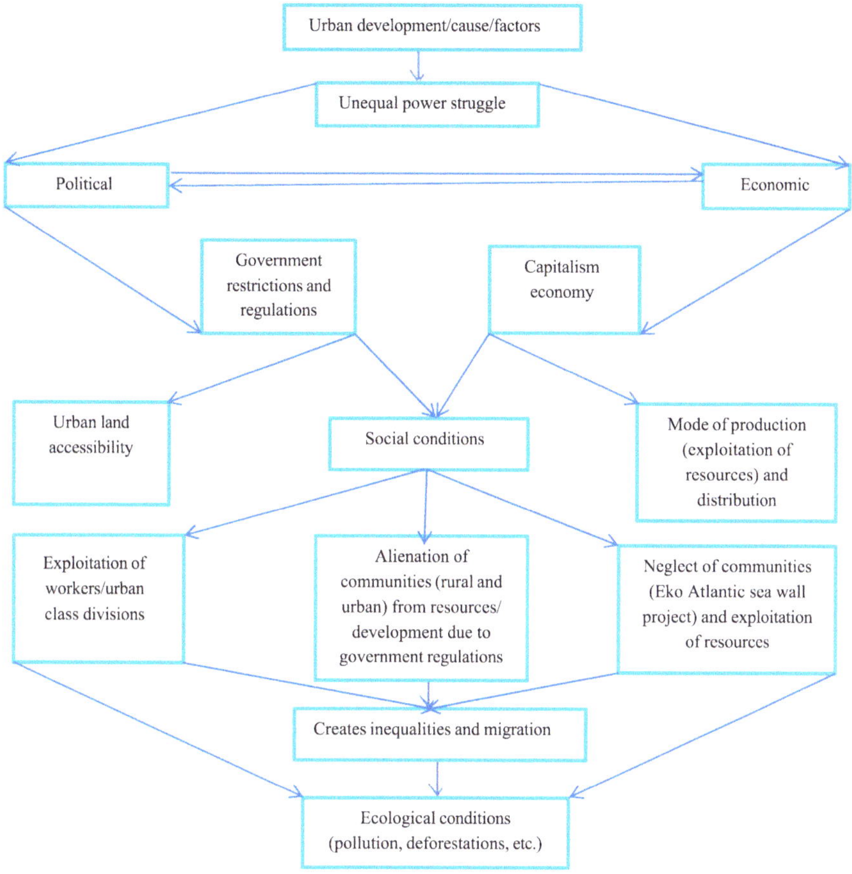

Fig. 17.1 Research theoretical framework. *Sources*: Ajibade (2017), Swyngedouw and Kaika (2014), Omotosho (2015), Carmody and Owusu (2016)

Chapter 18
Consequences of Urban Development in Nigeria

This chapter focuses on providing the context and data used in answering the research question. Secondary data, which was available for the period of 2000–2017 was retrieved from World Bank, World Atlas, Global Forest Watch, and Nigeria Data Portal. This will be used to answer research question by choosing key indicators like, urban inequility (socio-economic), land reclamation, flooding, water security, loss of biodiversity, deforestation, and pollution (environment). These indicators will help to show why the consequences of urban development in Nigeria. The Eko Atlantic city project and Makoko slum case studies will help to future explore the research question.

18.1 Socioeconomic Consequences

18.1.1 Urban Inequalities

Environmental factors and a case study of Eko Atlantic City (EAC) project and the Makoko Slum will be used to show the effect of urban development in Nigeria. It creates social inequalities whereby the poor cannot afford urban housing due to political and economic conditions and are therefore forced to live in slums.

Eko Atlantic City is built on land reclaimed from the Atlantic Ocean. It is located adjacent to Victoria Island on the North end, and along the coastline, off the Bar Beach. Commissioning of the dredging operations to create the foundation of the project took place in February 2008. This project is expected to consume 95 million m^3 of sand from the sea shelf of the Atlantic Ocean. Upon completion, it will be protected by a long stretch of wall comprising boulders and huge interlocking concrete blocks to form the "Great Wall of Lagos" (Eko Atlantic City 2012). The project will produce nine million square meters of reclaimed land for development and is expected to reduce coastline erosions that have plagued the coastline areas for

© The Author(s), under exclusive license to Springer Nature Singapore Pte Ltd. 2025
R. Perrault et al., *Sub-Saharan African Sustainable Development*, City Development: Issues and Best Practices, https://doi.org/10.1007/978-981-95-1458-8_18

years. The Bar Beach coastline is said to have eroded by at least 2 km in the last 100 years (Eko Atlantic 2012). This project is expected to accommodate 250,000 people and provide 150,000 commuters daily (Idowu and Home 2017). It was developed by Chagoury brothers, two Nigerian-born billionaires of Lebanese descent who own South Energyx Nigeria Ltd., a real estate development company created to build Eko Atlantic (Onuoha 2017). Eko Atlantic is meant to be an antidote to Lagos's deepest recession in 30 years. The Eko Atlantic Project also claims to have environmental assessment (EIA) needed for the oncoming project (Eko Atlantic City 2012), although in 2012, a storm surge in the exact area of Eko Atlantic's construction killed 16 people, and the aftermath of the tragedy was blamed on the disturbances caused by the Eko Atlantic building. The developers of the project highlighted however said that the development of "The Great Wall of Lagos" will protect the city from the effects of the surrounding Atlantic Ocean (Onuoha 2017).

As Ajibade (2017) explained, people from Kuramo beach, which is located beside the Bar Beach, have recorded human and livelihood losses since the commencement of land reclamation for the EAC. Many mentioned that they experienced low business sales and loss of lives due to increasing in storm severity and frequency, which they attribute to sand reclaiming activities for the project. A fatal tsunami of the Atlantic Ocean on August 18, 2012, displaced 1,500 residents in Kuramo and killed 16 people, mostly women and children. The devastation went beyond Kuramo to other coastal communities such as Akodo, Kaiyetoro, Alpha, Goshen Estate, and Lekki Peninsula. Kuramo locals saw this as dubious and expressed frustrations and anger over the protection of wealthy Lagosians with a massive seawall vis-a-vis their eviction in response to severe weather events. This dispossession of low-income residents from the beach areas and the appropriation of the common resource for elite estate development have served to deepen social inequality and livelihood vulnerabilities. Residents of Kuramo are now scattered around the city as they have nowhere to go, no means to rebuild their livelihoods, and no way to afford a place in the new development. By contrast, higher-income communities in close proximity to the ocean and lagoon were not displaced, and in some cases were given permission to expand beyond original buildable areas. Examples include Banana Island, Periwinkle Island, the EAC, and the newly proposed Orange Island. Viewed from a social justice perspective, the EAC as an adaptation project did not enhance the resilience and living conditions of the poor but in fact served to exacerbate their vulnerabilities. This transfer of collective property rights by the government into private hands may consequently lead to intensified class struggles, increased competition for ecosystem services, extensive commercialization of the ocean, and further extraction of urban surpluses (Ajibade 2017).

According to Dr Alan Blumberg a renowned researcher, what happens when you build a wall is that the approaching storm surge will come up to the wall, and it will move to the left or right of the wall, searching for a weakness or a low spot. If there are two islands close to each other, it may go to the island next door instead (Onuoha 2017). The wealthy and powerful may, in fact, take climate change seriously, not as a demand to modify their behaviors or question the fossil-fuel driven global economy that has made it possible, but as the biggest opportunity yet to realize unfettered accumulation and consumption (Lukacs 2014). The disaster capitalists behind Eko Atlantic have seized on climate change to push through pro-corporate plans to build a city that is not favoured by ordinary Nigerians. The poor abandoned outside its walls may once have served as sufficient justification for their flight and fortification, but now they have the very real threat of climate change as well (Lukacs 2014). The residential and business development, which will be home to a quarter of a million people and employ another 150,000, also has received its fair share of criticism. Promotional videos, images and virtual tours show a gleaming city of modern high-rises with a centre boulevard supplied with sustainable electricity, water, sewage, security and mass transit. But the impoverished residents of Makoko, who live just a few miles down from the project site in an aquatic slum will not benefit from Eko Atlantic's pricey apartments and clean energy. Neither will the millions of other Nigerians who live on less than one dollar per day (Winsor 2015).

There is no shortage of doubters and critics of the initiative, which is seen as an exercise in runaway neoliberalism by a country that cannot ensure 30 days of continuous power supply to its citizens (Winsor 2015). Makoko is a slum neighbourhood established in the eighteenth century as a fishing village with a population of roughly 100,000. A large part of the area rests in structures constructed on stilts above the Lagos Lagoon habituated by very poor urban dwellers (Ajayi et al. 2014) (Fig. 18.1).

The Makoko slums exist in their own peculiar world (Creighton 2014). Dubbed—with a heavy sense of irony—the Venice of Africa, the oily waters provide a way of life for the fishing community, but also the greatest dangers: spreading disease through the cramped population. Sprawling out from what was originally a small village in the eighteenth century, it has been consumed by Lagos, Nigeria's largest city. Guardian Lagos explained that Makoko is actually six distinct "villages" spread across land and water: Oko Agbon, Adogbo, Migbewhe, Yanshiwhe, Sogunro and Apollo. The first four are the floating communities, known as "Makoko on water";

Fig. 18.1 Makoko Slum. *Source*: *Ayorinde Ogundele (2022)*

the rest are based on land but both sides are united by the water, upon which they depend for their livelihood (Creighton 2014).

According to Adeola (2016) this slum settlement provides its residents with the shelter to live and work with the lagoon serving as the main source of livelihood from the sale of fish to the rest of the city of Lagos. However, the government of Lagos has unfortunately identified the growth of this neighbourhood as illegal and dangerous, as the unhealthy environment continuously expands, causing water contamination and flooding as a result of the lack of sanitation and waste management (Adeola 2016). With poor hygiene and sanitary conditions, lack of toilets, constant exposure to vectors borne diseases, and extreme weather conditions, these people are vulnerable to climate change and its rancid effects on human health. It is even scarier when one considers that Nigeria has over a thousand of these kinds of slums, with the bulk of them being in Lagos. With climate change wreaking havoc on the health of communities, these slum dwellers are at high risk of diseases, major epidemics and even death (Adebayo 2017).

As Balalobi (2013) explained, the Lagos state government regards Makoko as an environmental nuisance and a Water International 927 eyesore. Government officials are interested more in demolishing the community's "illegal structures" than in upgrading them. Officials of the state government have demolished some of the makeshift canal toilets, as they are viewed as hindering the free flow of wastewater and storm water in the canal. The demolition of the toilets further compounded the sanitation crisis in the community, as the only option left for residents of all ages is to defecate openly and directly into the lagoon that lies underneath their makeshift houses. Open defecation is the main means of "sanitation" in the community, and the prevalence of this practice has adverse effects on the health of the residents (Babalobi 2013). The court put a stay on future evictions and initiated a process to mediate some kind of agreement between residents and the government. Nevertheless, the police soon returned with a crew of demolition workers and bulldozers, this time with a mission to "clear" structures "to ensure that the waterfront area is free from environmentally injurious and unsanitary habitation." Another 4,700 residents lost their homes (Mcdonnell 2017). An argument in support of the slum clearing action by the government was presented as an effort to clean up Lagos, and ensure environmental sustainability, as the unwholesome structures constitute an environmental nuisance, security risk, an impediment to the economic and gainful utilization of the waterfront (Adeola 2016). Demolitions aren't a solution to slums according to a Lagos resident, they just push people deeper into poverty, and they move to a different slum (Mcdonnell 2017).

According to Neuwirth, not everyone is a fan of slum redevelopment or plans for regeneration because of the idea of deliberately transforming slums into models of urban development. "Why can't communities simply be communities and develop in the organic way that we allow other communities to develop?" Pointing to Brazil's favelas and Istanbul's gecekondu communities, he says: "They are inspirational in that people have developed them themselves, without government and real estate types pushing them around. Without a doubt, they still have problems. But they are stabilising themselves and, over time, knitting themselves into the fabric of their cities. This is a true marvel of global urbanism (Ogunlesi 2017).

Safety nets for the poor and displaced in the country are so lacking that, months later, many of those made homeless by this assault are still living in their boats, open to elements of floating on the water (Creighton 2014). With the increase in downpours, floods and rising sea levels, these slum dwellers in Makoko, whose existence is almost entirely on water and whose shelters are ramshackle combinations of wood and rusted zinc, are at an alarming risk of displacement as their shelters are not built to adapt to the changing climate. Displacement means more vulnerability, more climate displaced persons, more mental health disorders and more spread of diseases (Adebayo 2017).

Makoko's biggest health challenges are not communicable diseases such as cholera, but non-communicable diseases: malaria, respiratory diseases and malnutrition (Ogunlesi 2017). With the extreme rising temperature and floods, the water becomes breeding ground for mosquitoes, leading to malaria for most of the population which has little access to medical supplies without a standard hospital, facilities, equipment and available health personnel in the area. In addition, reports by Adeola, (2016) also identified Makoko as a slum that is easily visible from the Third Mainland Bridge which connects Lagos mainland to the city's rich and business districts, and its scenery does not represent Lagos as a true emerging world-class city in positive light. The case of Makoko substantiates Lefebvre's spatial Marxism, where capitalism reaches globally, so deep into the earth while creating a conflict of interest between economic advancement and the everyday life realities on the other. Lefebvre argues that an authentic knowledge of space is required in the question of its production, this as a major supporter of the critique of the Nigerian urban space; most importantly as globalization has taken the centre stage over local action. The original residents of Makoko are today subjected to forced eviction, demolition and neglect, as the big price for the modernization of the city of Lagos (Adeola 2016). Makoko is an area of practical urban inequalities and social depreciation, where open defecation is a lifestyle with no social value to toilets, sewage is indiscriminately disposed and virtually no cholera risk awareness (Ayeni 2014).

As Carmody and Owusu (2016) emphasize, neoliberalism has significantly curtailed the role of the state in Africa, and along some dimensions, incapacitated and weakened the planner's toolkit for dealing with the problems of African cities. In addition, the focus of urban planners across Africa seems to have shifted from making the city a decent place to live for its residents, to a place of elite consumption and production which has resulted in the pervasive lack of urban amenities, widespread poverty and the resultant slumification of African cities (Carmody and Owusu 2016). This is also used to explain the Eko Atlantic City project which has been built for the rich Nigerians, and the Makoko slum where the majority of the poor income earners live. As explained by Lukacs, (2014) Eko Atlantic is where you can begin to see a possible future a vision of privatized green enclaves for the ultra-rich ringed by slums lacking water or electricity, in which a surplus population scramble for depleting resources and shelter to fend off the coming floods and storms. Protected by guards, guns, and an insurmountable gully of real estate prices the rich will shield themselves from the rising tides of poverty and a sea that is literally rising. A world in which the rich and powerful exploit the global ecological crisis to widen and entrench already extreme inequalities and seal themselves off from its impacts: this is a climate apartheid (Lukacs 2014).

18.2 Environmental Consequences

18.2.1 Soil Degradation/Land Reclamation

Land reclamation from seas and wetlands around the world is not entirely new. It entails dredging and transporting a large amount of sand from the seabed to create artificial land for infrastructural and industrial development (Idowu and Home 2017). This seabed is mostly inhabited by aquatic species, which when sand mining takes place, are dangerous to aquatic lives. Besides building constructions within the gazetted setbacks and within the floodplains, another construction-related factor identified was the hydrological modification of streams and rivers within Nigeria cities through diversion and land reclamation activities by filling river courses with sand to gain additional land space on which to build. The natural courses of rivers and streams have been diverted using various manmade structures. Areas that were initially floodplains have been sand-filled and transformed to accommodate residential, commercial, and religious land use (Agbola et al. 2012).

In the Lagos metropolis in particular, 1.31% of the total water body in Lagos within the period of 1990 to 2006 has been lost to the collective efforts of continuous land reclamation and sand filling of wetland to cater for the need for residential development either by the government or private developers (Nwokoro and Dekolo 2012). A good analysis is the Tinubu Administration in July 2006 which awarded South Energyx Nigeria Limited (SENL) the exclusive rights to reclaim land, build infrastructure and oversee the development of the EAC[1]. It also signed a global Certificate of Occupancy under a 78-year lease agreement with the company. Sand dredging from the marine shelf waters to reclaim 10 million square meters of land in the Bar Beach area began in 2009. By 2013, well over 5 million square meters of land had been reclaimed and the development of this land is currently underway. When completed, a five-mile sea revetment called the 'Great Wall of Lagos' will be built around the city to protect it from rising sea levels and flooding and simultaneously act to prevent further erosion of the coastline (Ajibade 2017).

Agboola and Ayanlade's (2016) findings imply that the urban dwellers settled along the coast and developments might be helpless against the rise in sea-level in Ikorodu and Alimosho[2] especially due to the construction of the seawall (Agboola and Ayanlade 2016). The manifestations of the weak institutional frameworks are

[1] [24] EAC is a private funded project on reclaim land from the ocean in Lagos state.

[2] Ikorodu and Alimosho are areas which are closer to the Atlantic Ocean and also populated by urbanites in Lagos state.

reflected in many detrimental activities taking place in the city such as building on river floodplains and indiscriminate waste disposal. This weakness of existing institutional frameworks for the monitoring and management of the urban environment have also contributed to the flooding issues in urban cities (Agbola et al. 2012).

18.2.2 Flooding

Flooding is also of the consequences of unplanned urbanization in Nigeria. Flooding in various parts of Nigeria has occurred more frequently in recent years because of the escalating trend towards urbanization and the increasing number and size of metropolises and mega- cities without proper planning and drainage infrastructure (Adeloye and Rustum 2011). Uncontrolled urban population growth associated with poor governance might have encouraged the decay of existing infrastructure and hamper the development of adequate infrastructure in many cities in Nigeria. This has made flooding an unavoidable security risk to many urban centres in Nigeria (Akiyode 2013). This also shows that alteration in land cover and poor watershed management particularly in floodplains of rivers are an important issue for consideration. This is because infiltration capacity of such land surfaces is affected thereby increasing runoff rates and exacerbating flooding from high rainfall which has major consequences for the environment and human population (Adelekan 2011).

Floods have serious negative impacts on the socio-economic and health conditions of the population. In 2012, thousands of people were displaced from their homes, properties were destroyed and lives lost due to unusual extreme flooding. The most affected were the poorer neighbourhoods of the islands, wetlands and low-lying areas of the state and it is important to note that these climate-induced extreme events are increasing in frequency and intensity (Idowu and Home 2017). A good example of the effect of flooding in Nigeria was the issue of flooding in Lagos State in July 2017. A heavy rain that fell over the course of one night and morning left hundreds of homes and roads flooded in Nigeria's affluent neighbourhoods of Ajah, Ikoyi, Lekki and Victoria Island (Bukola and Ige 2017), other areas included the Victoria Garden City (Channels Television 2017).

Lagos residents who spoke about the situation on condition of anonymity said that the major issue is the not the heavy rain in Lagos. It is the fact that over the years, there was no proper planning of the city by successive governments, despite the rapid urbanization and population increase (Channels Television 2017). According to Bukola and Ige (2017), Adelakun who specializes in climate, development and human dimension of the global environmental change said that the flooding in Lagos was due to heavy rain and high population. Natural surfaces, like soil and vegetated area absorbs the water when it rains and infiltration can take place, but when you have artificial surfaces such as (construction, roads and buildings), there is more run-off. Also, in many parts of Lagos the drainage capacity is low. In some cases, the drains were constructed many years ago. Now, there is more and heavier rainfall, the capacity of the drainage system is not enough to carry away the stormwater and a lot of the drainage channels are blocked with a lot of people depositing water sachets and tyres among other things (Bukola and Ige 2017).

Adelakun also talked about the government encouraging people to allow space for natural areas when building. Adelakun also gave another reason for the flooding: loss of land. Lagos areas are wetlands, and a lot of land reclamation, especially in terms of sand filling, is taking place. Wetlands are buffers for floods, so when sand is filled into swampy areas to make way for buildings; there is also a reduction in the capacity of the natural environment to serve as buffer for flood (Bukola and Ige 2017). Another good example is the flooding in Benue.

State[3] in August 2017 which displaces thousands of people, loss of property and livelihood after a heavy rainfall.

In the early hours of August 27 2017, a torrential downpour caused the River Benue to breach its banks across the marshy plains around Makurdi,[4] washing off a deluge of muddy water into thousands of homes overnight. The problem was compounded by rainwater from the upper side of the town that poor drainage system hindered from finding its way through the swamps into the River Benue, trapping in the low-level areas (Ogundipe 2017).

Apart from rendering thousands homeless, the disaster also washed away villages, farmlands and food storage facilities among others. The first step, according to disastermgmt.org, is to keep the drainage system clean. This allows water to be carried down quickly. Choked drains cause a significant reduction in the ability and speed of the water to be drained away. In most situations of urban flooding, this is a major cause. The drains might choke due to the throwing of solid waste inside storm drains. These solid wastes might include construction material, plastics, paper etc. This is a clear example of how human activities can amplify the process of flooding (Okolie 2017).

A statement released by State Emergency Management Agency (SEMA) in 2017 disclosed that in Achusa, about 200 houses were affected with 5,125 persons displaced. In Idye, 217 houses were affected while 5,200 persons were displaced. Also, behind the Civil Service Commission, 200 houses were submerged in flood and 5,777 persons displaced. At Genabe, 200 houses were also affected with 5,021 persons displaced; 218 houses around the Wurukum market were affected with 1,000 persons displaced, and at Wadata market, 150 houses were affected and 4,300 persons were displaced. At Industrial Layout, 69 houses were flooded and 4,310 persons displaced; in Demekpe, 111 houses were affected while 7,820 persons were displaced. The statement further revealed that 137 houses were flooded in Katungu with 6,031 persons displaced while at Agboughul-Wadata, 201 houses were affected and 5,728 persons displaced. Although two camps were set up in Makurdi to accommodate those rendered homeless, a total of 4,776 people have been registered from 546 households. However, cases of malaria, typhoid fever and diarrhea have been reported in the camps (Okolie 2017). The situation also received a slow response from the National Emergency Management Agency (NEMA) on the disaster (Okolie

[3] Benue state is a state in northern part of Nigeria.

[4] Makurdi is a city in the northern part of Nigeria.

2017). NEMA, an agency was created by the federal government with the mandate to respond speedily in times of emergency.

As CBD explains, flooding is aggravated by poor land use and watershed management practices. Human activities such as unplanned urbanization, blockage of river/drainage channels, land clearing for agricultural purposes and deforestation contribute immensely to flooding (CBD 2016). A study of urbanization trends in Kaduna city in the Northern part of Nigeria over a period 1962 and 2009 indicated that increasing urbanization along the Kaduna River floodplain is responsible for the problem of flooding that has been experienced in recent times along the river floodplain (Akiyode 2013).

18.3 Water Security

Reclamation of lands in Nigeria for urban constructions, which have increased flooding, also affects the availability of freshwater. As Idowu (2017) explains, this is because of the salinization of the groundwater. In places such as Victoria Island, Lekki and Ikoyi, it is not uncommon to have drinking and potable water being supplied through water tankers due to water pollution (Idowu and Home 2017).

One of the causes of groundwater pollution of saltwater is due to the annual increase in sea levels (Idowu and Home 2017). One of the resultant effects may be damage to water storage and supply systems, and saltwater intrusion into groundwater resources in the coastal areas. These pose a further threat to the availability of drinking water not just in these coastal areas and wetlands, but also landward areas (Idowu and Home 2017) which can lead to water scarcity in Nigeria. This water scarcity will create the tendency to concentrate populations around the remaining limited sources of water, and under such circumstances, there is increased possibility of additional contamination of the limited sources of water and transmission of water-borne diseases like cholera, typhoid fever, guinea worm infections and river blindness (Odjugo et al. 2010).

As per UN-Habitat (2017), adequate water supply is essential for environmental sustainability and quality of life. Access to clean water reduces morbidity and mortality and improves the productive abilities of the poor. Water scarcity characterizes African cities. Although official statistics reveal that 89% of the urban population in Africa is now enjoying improved water supply, a large majority of Sub-Saharan African cities experience regular water shortages. The UN-Habitat 2017 survey shows that 11 of the 14 African cities (79%) are associated with such serious problems. Experts concur that areas affected the most are Ibadan and Dar es Salaam, closely followed by Accra, Addis Ababa, Luanda, Lusaka, Lagos and Nairobi, which suffer continuous water shortages. Despite 78 to 98% of households in four of these cities (Accra, Lagos, Nairobi and Lusaka) benefiting from improved access to water, most experts also mention them as experiencing severe water shortages (UN-Habitat 2017).

Eni et al. (2011) state that urban development affects the quality and quantity of underlying sub-surface water by radically changing patterns and rates of recharge by initiating new concept regimes and adversely affecting groundwater quality. An assessment of the risk to groundwater from urban processes needs to take into account the interaction between the recharge and discharge pressures and the pollutant loading on the one hand, and the nature of the subsurface environment on the other. The potential for urbanization processes to have an impact on the underlying groundwater is a function both of the aquifer's vulnerability to pollution and its susceptibility to the consequences of excessive abstraction (Eni et al. 2011).

Another factor is the Lake Chad Basin which serves as a source of freshwater for over 20 million people who depend on it for their livelihoods. The Lake, with its rich biodiversity is a valuable resource source for the fishery industry, and pastoral and agricultural activities for the inhabitants in the Basin. However, the lake has continued to shrink with desertification. It is estimated that between 1963 and 2007, the surface area of the lake has shrunk from 25,000 km^2 to less than 3,000 km^2 with little or no water in the Lake in the Niger and Nigerian portions (Babanyara et al. 2010). Moreover, freshwater availability has become one of the most critical factors in the country's development. Increasing numbers of Nigerians are now dependent on groundwater as their primary source of freshwater. Inappropriate management of freshwater and competition between user groups limit efforts by the government to develop the nation's economy and improve the standard of living of the citizens. There are also concerns over freshwater quality, in terms of pollution from domestic effluents and industrial wastewater, particularly in the coastal zone of Lagos and the oil-producing Niger-Delta region (CBD 2016).

A study by Ayeni (2014) shows the results of sampled water from Makoko, which revealed that 40% of the 10 water samples collected from the stilt houses on Lagoon and 33.3% of the 6 water samples collected from over-land residents were confirmed to contained bacterium Vibrio cholera presence. The presence of fecal coliform in the sampled water also suggests high contamination of sewage and animal waste. It appeared that most probable incidences were caused by inadequate drinking water sources (Ayeni 2014).

In terms of water pollution, the World Health Organization (WHO) reports that there are serious problems in Nigeria, which threaten the health of both rural and urban populations. In rural areas, water pollution occurs mainly because of agricultural activities. In urban and peri- urban areas, "many factories are located on riverbanks and use the rivers as open sewers for their effluents. The petroleum industry represents the greatest threat to water quality. In the Niger Delta, in particular, water pollution associated with the oil industry has resulted in such extensive oxygen depletion in some bodies of water that aquatic fauna died of asphyxiation. Such pollution also threatens the ecosystems of Nigeria's mangroves (Babanyara et al. 2010). Waste enters these bodies of water in both solid and liquid forms. These are mostly derived from industrial activities. As a result, bodies of water which are major receptacles of treated and untreated or partially treated industrial wastes have become highly polluted. The negative results of this on public health and the environment are usually great in magnitude. These include endangering aquatic resources and other commercially important marine flora and fauna (Osibanjo et al. 2011).

18.4 Loss of Biodiversity

Biodiversity plays several important roles in urban environments (McKinney 2008). These roles include ecosystem services such as air and water purification, and amenity values such as aesthetic enjoyment and recreation. In addition, urban biodiversity has an important role in educating an increasingly urban population about nature and species conservation. On one hand, some aspects of urbanization promote the loss of species diversity; on the other, the negative impact on biodiversity is a simplification of vegetation in many areas. Landscaping and maintenance of residential and commercial areas typically involve removal of shrubs and dead wood and an increase in grasses and herbs. This has a negative impact on the diversity of birds and other animals, whose diversity tends to correlate with vegetative complexity and plant species richness (McKinney 2008).

Additionally, there has been destabilization of urban biodiversity in most of the urban communities in Nigeria. Land cover, land use and urban space in most cities have been changed due to urbanization that has triggered infrastructural development and societal change. This leads to both terrestrial and aquatic biodiversity loss in many of the urban centres. 11% of the bird species in the urban areas in Nigeria may be threatened by developmental activities within urban settings (Akiyode 2013).

However, species extinction is a natural process according to Rockström et al. (2009) that would occur without human actions, however, biodiversity loss in the Anthropocene has accelerated. Species are becoming extinct at a rate that has not been seen since the last global mass-extinction event. The fossil record shows that the background extinction rate for marine life is 0.1–1 extinction per million species per year; for mammals, it is 0.2–0.5 extinctions per million species per year. Today, the rate of extinction of species is estimated to be 100 to 1,000 times more than what could be considered natural (Rockström et al. 2009). This is because human activities have increased over the years, especially in consecrated spots being in urban settings and cities.

Urban development is a new selective force that is changing the composition of animal communities tremendously. During the process of urbanization, some animal species will disappear from the newly urbanized habitat. For example, birds that feed on specific natural food sources will fly to other areas to search for food. Species that cannot move as fast as urbanization is expanding will disappear. But in contrast to this other species, such as crows and doves, will move into the city or increase in density. In this way urbanization also filters bird communities (Isaksson and Sumasgutner 2016). Natural and human-made threats, socio-cultural problems, as well as direct and indirect consequences of socio-economic development, have contributed to the erosion of biodiversity at all levels. Within the last 30 years, about 43% of the forest ecosystem has been lost through human activity. Close to 96% of the original 20%

forest cover has been cleared, and only 2% of what remains is undisturbed. This has resulted in massive loss of the very rich and diverse forest flora and fauna, including indigenous tree species. Large parts of the savannah are also being ravaged by human-induced desertification, causing ecosystem impoverishment and biodiversity loss. The protected area system in Nigeria currently covers only 5.7% of the landmass. The number of threatened and endangered species has also been on the rise. It is estimated that 0.4% of the plant species are threatened and 8.5% endangered, with 0.14% of the animals and insects threatened and 0.22% endangered. In addition, about 10–12 species of primates are threatened (CBD 2016).

A land use and vegetation cover study by Forest Management Evaluation and Coordinating Unit (FORMECU) of Federal Department of Forestry, for the periods 1976/1978 and 1993/1995 showed that the changes in vegetation during the interval of 15 years (1978–1995), was staggering, with 32% of the riparian vegetation cover removed by the year 1995. Forest cover became less than 20% that of 1978, and it is presently estimated that forest cover for Nigeria is less than 10%, with increased distribution of alluvial and rock outcrops. With increased population, it is not surprising that there would be intensive agricultural land use. It is presently estimated that Nigeria is losing some 351,000 km^2 to the desert, representing 40% of its total land mass due to desertification effect. In Nigeria alone, it is estimated that more than 38 million people, representing more than 32% of the total population of the country, live in areas subjected to desertification (Babanyara et al. 2010).

18.5 Deforestation

Another means by which species are endangered is deforestation, which is one of the effects of urban development. As per Babanyara et al. (2010), deforestation is the process whereby trees are felled for several purposes, but without replanting to replace the one chopped. The continuous removal or destruction of significant areas of forest cover results in a highly degraded environment with an attendant reduction in biodiversity. It also causes soil erosion, and marginal lands can lead to desertification. Nigeria probably has the world's highest deforestation rate of primary forests today, having lost more than half of its primary forests in the last five years (Babanyara et al. 2010).

Deforestation is also the major reason for the conversion of forested land for cities. A lot of exotic species are now on the verge of extinction because of this and loss of habitat for local flora and fauna (Khan 2017). A good example of this is the construction of the Superhighway around the remaining largest tropical rainforest by the Cross River Government in Calabar,[5] one of the most populated cities in Nigeria. According to Adeniyi Karunwi, Director General of the Nigerian Conservation Foundation (NCF), the Cross River Tropical Rainforest is the largest remaining rainforest ecosystem in Nigeria and one of the top 25 biodiversity hotspots in the world facing a

[5] Calabar is a city in southern Nigeria.

very serious threat of degradation on account of the proposed Superhighway project of the Cross River State Government (Offon 2017).

This rich forest is serving as home to highly threatened species including 1,568 plant species, of which 77 are endangered medicinal plants and orchids, 22 primate species such as the Cross River gorilla (with only about 300 individuals remaining in the wild), Nigeria-Cameroon chimpanzee, drill monkeys, and a range of other wildlife species facing one of the greatest conservation challenges of our time (Offon 2017). Environmental activists have, however, kicked against the project which they say would destroy several species of plants and animals, and also displace the indigenous people of the areas (Ukpong 2017). The highway, which will cut across at least 16 local government areas when completed, is one of the signature projects of the administration of the Cross River State Government (Ukpong 2017).

According to Unaegbu (2017) the proposed Cross River Superhighway project is one of the most publicly contested developments in Nigeria. It has been contested by indigenous people, concerned citizens, and local and international organizations for many reasons. First was for the now reverted 20 km right of way, second, it goes round the Cross River National Park and a number of community forests, third the legitimacy of its funding source is in question, fourth, there has been much falsified information in the submitted Environmental Impact Assessment (EIA) reports, Fifth its lack of consideration and consultation of the Environmental Impact Assessment Decree No. 86 of the 1992 and of the affected communities (Unaegbu 2017). It is pertinent that the proposed Superhighway has already produced three failed reports. These reports have been widely criticized as being substandard, fraudulent and non-compliant with National Laws and International best practices according to (Offon 2017).

It is important to note here that the project has commenced in violation of the Land Use Decree and EIA Decree by Cross River State Government on February 20, 2017, who issued a notice to the Federal Government of Nigeria about going ahead with the Superhighway with or without the approval of the EIA (Unaegbu 2017). This action is aggravating widespread poverty and ultimately resulting in untold hardship among indigenous communities that live adjacent to these forests. Essentially, this unplanned and unsustainable developmental action of government pertaining to habitat destruction as well as ecological and species destabilization is generating deep environmental and socio-economic concerns among civil society groups, local communities and other stakeholders (Offon 2017). The graph (Fig. 18.2) below shows the rate of deforestation in Nigeria.

Deforestation has caused many environmental and social problems. From 2001 to 2017, Nigeria lost 738kha of tree cover, equivalent to a 7.3% decrease since 2000, and 56.3^{Mt} of CO^2 emissions (Global Forest Watch 2017),thereby leading to biodiversity loss (species threatened: birds 21,fish 71, mammal 29 and plants 197) (Knoema 2016) and destruction of local livelihood. The graph above shows that from 2001 Nigeria has lost, 43,386 Ha to 171,000Ha in 2017. According to Global Forest Watch, 37.5% of Green House Gas (GHG) emissions in Nigeria came from land-use change and forestry in 2011 and in 2010 Nigeria had 8.86Mha of tree cover, extending over 9.7% of its land area. In 2017, it lost 170kha of forest, equivalent to

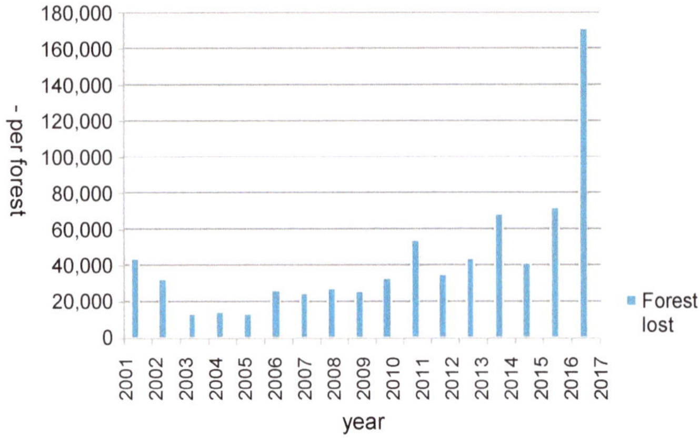

Fig. 18.2 The rate of deforestation from 2001 to 2017. *Source*: Global Forest Watch (2017)

12.1^{Mt} of CO^2 of emissions (Global Forest Watch 2017). About 11,089,000 hectares of Nigeria is forested as explained by Mongabay (2006), of this, 2.9% or roughly 326,000 hectares is classified as primary forest, the most biodiverse form of forest. Between 1990 and 2000, Nigeria lost an average of 409,700 hectares of forest per year. This amounts to an average annual deforestation rate of 2.38%. Between 2000 and 2005, the rate of forest change increased by 31.2% to 3.12% per annum. In total, between 1990 and 2005, Nigeria lost 35.7% of its forest cover, or around 6,145,000 hectares. Nigeria lost 1,230,000 hectares of its primary forest cover during this time. Deforestation rates of primary cover have decreased by 79.1% since the end of the 1990s. Measuring the total rate of habitat conversion (defined as change in forest area, plus change in woodland area, minus net plantation expansion) for the 1990–2005 interval, Nigeria lost 39.2% of its forest and woodland habitat (Mongabay 2006). Nigeria has the world's highest deforestation rate of primary forests according to revised deforestation figures from the Food and Agriculture Organization of the United Nations (FAO 2017). Between 2000 and 2005 the country lost 55.7 percent of its primary forests.

Deforestation was also identified as an important contributing factor to the incidence of flooding in Ibadan. The increasing urbanization has resulted in the conversion of areas used for small- and medium-scale agriculture or forestry into building sites. The situation was compounded by the increasing deforestation of the wetlands and riparian forest areas including the Eleyele watershed, which served as storage areas for floodwater. Most of these wetlands and riparian forest areas have been cleared and laid out for residential development, such as is seen in the Apete community (Agbola et al. 2012). According to the Biodiversity Centre, development is meaningful if it does not increase a country's vulnerability to environmental impacts. If a nation's environmental foundations are depleted, its economy will decline, its social fabric may deteriorate, and its political structure may become destabilized. The environment is, however, a complex and interactive system consisting of the atmosphere, land surface and bodies of water, as well as living things. The degradation of a component of the ecosystem will have positive or negative feedback effects on the others (CBD 2016).

18.6 Pollution

Environmental pollution is rapidly increasing in Nigeria. The major causes of the increase include poor sanitation, inadequate solid waste disposal, effluent discharge, rapid and unplanned urbanization, deforestation, mining, and increasing use of chemical fertilizers and insecticides. Surface runoff collects all types of pollutants which are then moved into the rivers, dams and sometimes wells (Babanyara et al. 2010). Modern agriculture according to Rockström et al. (2009) is also a major cause of environmental pollution, including large- scale nitrogen- and phosphorus-induced environmental change. At the global level, the additional amounts of nitrogen and phosphorus created by humans are now so large that they significantly perturb the global cycles of these two important elements. Human processes — primarily the manufacture of fertilizer for food production and the cultivation of leguminous crops — convert around 120 million tonnes of N^2 from the atmosphere per year into reactive forms —which is more than the combined effects from all of Earth's terrestrial processes. Much of this new reactive nitrogen ends up in the environment, polluting waterways and coastal zones, accumulating in land systems and adding a number of gases to the atmosphere. It slowly erodes the resilience of important Earth subsystems. Nitrous oxide, for example, is one of the most important non-CO^2 greenhouse gases and thus directly increases ultraviolent radiation exposure (Rockström et al. 2009).

Apart from landfill gases that are readily sent into the atmosphere of residential areas in most of the urban centres in Nigeria, there are uncontrolled gas emission from the industries and factories in many of the industrial urban centres like Lagos, Kano, Ota, and Aba. A study by Akiyode (2013) finds that the urban environments of Nigeria are characterized by increased particulate matter (PM^{10}) The study further suggested that that the urban corridors of over 70% of Nigerian cities are sites having a high rate of daily mean/annual mean ambient particulate matter of over 120 μg/ m^3, while less than 30% of Nigerian urban centers had mean annual ambient particular matter value of 119.2 μg/m^3. Uncontrolled air pollution has socioeconomic, physical, chemical and health implications on cities and urban dwellers (Akiyode 2013). Air pollution is one of the key contributors to disease and fatalities globally and is a significant problem in urban areas in Nigeria. The main sources are industrial plants, residential and commercial buildings, concentrations of hazardous waste and motor vehicles (Gbadegesin et al. 2011).

It is pertinent to state that increasing health challenges is a critical issue in the analysis of city space. Most of the health hazards in the city are not dissociable from the increasing level of pollution. Constant emission of carbon monoxide and other unhealthy forms of biological and chemical substances into the atmosphere endangers the lives of people in both cities and their corresponding slums. Incomplete combustion of vehicular engines is not excusable in this regard. Similar to the problem of pollution is the issue of congestion which comes with poor ventilation. Poor ventilation in turn negatively impacts healthy respiration. As a result of the ever-increasing demand for housing, and inadequate planning, a good number of houses in Lagos are poorly ventilated (Akanle and Adejare 2017).

One of the most critical problems of Nigeria is the improper management of the vast number of wastes generated by anthropogenic activities. More challenging is the unsafe disposal of these wastes into the ambient environment. Bodies of water, especially freshwater reservoirs, are the most affected. This has often rendered these natural resources unsuitable for both primary and/or secondary usage. Industrial effluent contamination of natural water bodies has emerged as a major challenge in developing and densely populated countries like Nigeria. Estuaries and inland bodies of water, which are major sources of drinking water in Nigeria, are often contaminated by the activities of the adjoining populations and industrial establishments (Kanu and Achi 2011). The UN-Habitant (2017) explains that traffic congestion is a major environmental problem, and apart from the attendant air pollution, the cost to the local economy is enormous. Similar opinions are held by experts in Lusaka, Accra, Algiers, Lagos, Ibadan and Luanda, where rapid urban sprawl and uncontrolled spatial development combine with poor infrastructures and weak regulatory frameworks to undermine the quality of life and, more generally, prosperity in their respective cities (UN-Habitat 2017).

This is also gradually becoming a serious problem in Nigerian cities. Especially in the metropolitan areas, inefficient energy combustion in the transportation system generates high levels of localized air pollution. The recent increase in the importation of second-hand cars and the widespread adoption of the single engine "Okada" motorcycles for ferrying passengers all over most Nigerian cities, have accentuated the general level of air pollution (Babanyara et al. 2010). Traffic congestion, which increases the generation of the automobile exhaust due to the traffic delays, is another major environmental problem in Nigerian urban centres (Oyeleye Oyewale 2013).

The graph below illustrates the amount of greenhouse gas per thousand tons omitted into the atmosphere through deforestation, traffic congestion, industries etc. The data below explains pollution gotten from greenhouse gases and its fluctuations from 2001 to 2012. According to World Bank (2016), other greenhouse gas (hydro fluorocarbons, per fluorocarbons, and sulphur hexafluoride) emissions are relatively small and more dangerous than carbon dioxide, with much higher atmospheric lifetimes and high global warming potential. Methane is produced from human activities such as agriculture and from industrial methane production. Nitrous oxide is mainly from fossil fuel combustion, fertilizers, rainforest fires, and animal waste and while CO^2 emission is a naturally occurring gas fixed by photosynthesis into organic matter. A by-product of fossil fuel combustion and biomass burning it is also emitted from land use changes and other industrial processes. It is the principal anthropogenic greenhouse gas that affects the Earth's radiative balance (Fig. 18.3).

These greenhouse gases have been fluctuating over the years, the CO^2 in 2002 was 76,057.2 and in 2012 were 98,503. CO^2 emission was at the highest from the

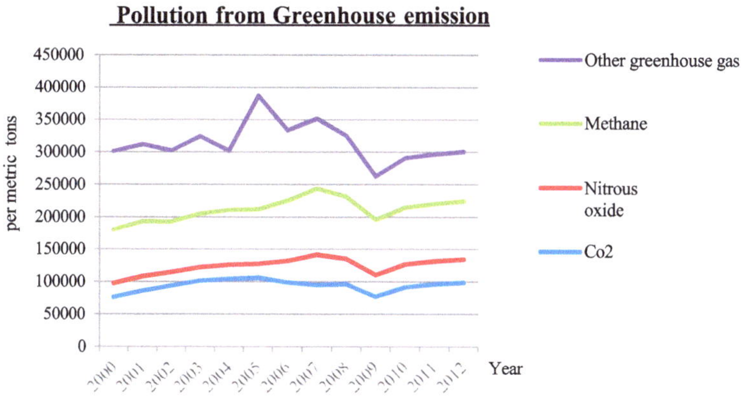

Fig. 18.3 Pollution from Greenhouse Gas emissions. *Source*: World Bank Data (2016)

period of 2003 to 2006 (101,616.2–98,891.7) when Nigeria lost 55% of its forest according to FAO (FAO 2017). Nitrous oxide rose from 21,186.1 in 2000 to 36,184.7 in 2012. Methane rose from 82,741.6 in 2002 to 89,781.9 in 2012 while other greenhouse gases have reduced from 121,237.1 in 2002 to 76,218.8 in 2012. According to CBD, the atmosphere is vital for the survival of man and other living animals. It provides air for respiration and photosynthetic processes. It also provides a safe environment surrounding man and other living organisms by shielding them from dangerous particles and rays. It is also the habitat for varied flying organisms and the medium for air navigation, but the problem of air pollution in Nigeria cities is becoming devastating (CBD 2016).

The study shows that people migrate due to economic condition to improve their livelihood and have a better life. This movement to urban cities is however not controlled due to inadequate planning by the government to provide basic necessities, especially for the rural communities. These poor government policies are also explained in the construction of the Eko Atlantic City project, especially in relation to the building of a seawall and violation of the environmental assessments to build a city, which won't be affordable to the average Nigerian. This violation is also seen in the construction of the superhighway, without making necessary accommodations for rural communities who live and earn from this forest. The Makoko slum is the epitome of another failure of the government to provide allocation equally and making resources available to urban dwellers in Nigeria. This illustration boils down to the conceptual model and urban theoretical framework on how the environment becomes a tool in the court of both political and economic conditions which feeds on the marginalization of the lower class. The study also explains the environmental and socioeconomic consequences of rapid urbanization in Nigeria.

The economic conditions affect the ability or the power of the government to make laws and regulations which typically are not favourable to the urban masses. This also leads to exploitation of workers, alienation/neglect of communities, and the shift of focus away from the development of rural communities, thus leading to migration to urban centres. The environment is affected due to the economic exploitation of resources and the marginalization of the poor masses since they too exploit the environment in order to survive.

It is important to note that these theories present a clear view that accompanying urban socio-economic restructuring also contributes to changes in the ecological forms of urban areas more broadly. While environmental, (both social and physical), qualities may be enhanced in places, for some people, they often lead to a deterioration of social and physical conditions and qualities elsewhere, both within cities and between cities (Heynen et al. 2006). The UPE theory has shown a comprehensive framework for studying human–environment interactions and the uneven urban environmental development that affect human society in both beneficial and harmful ways (Roy 2011). As Zimmer (2010) explained, the understanding of metabolism in Urban Political Ecology is also rooted in a Marxist theory that places human labour at the centre of it. Whereas eco-socialism takes the term as denoting something purely material such as the flow of energy and matter, Urban Political Ecology takes a different view. Based on Marx, metabolism in Urban Political Ecology is taken as a material or energetic exchange, but this exchange is seen as a historical product. This means that humans are able to control their input according to intention and interest so that the metabolism is the result of specific drives, desires, and imaginations. The aim of humans therein is the satisfaction of their respective needs. However, not all humans can reach this aim equally the reason for this being that the metabolism is mobilized and realized within existing social relations. The metabolism is designed in such a way that surplus values produced by it go to certain social classes in which the elites get a bigger share of its profits (Zimmer 2010).

Escobar (2006) also emphasizes that the value of nature cannot be assessed only in economic terms. There are ecological and political processes that contribute to defining the value of natural resources which cannot be reflected in market prices. Indeed, in many cases, there is incommensurability between economic and ecological processes to the extent that communities value the environment for reasons other than economic, for example, when they consider nature to be sacred and uncommodifiable. Conflicts over access and control of resources take on a complex ecological and political character if the widely held idea that everything can be tagged in monetary terms is suspended. This is why ecological economists have suggested the category of ecological distribution. Under conditions of unequal distribution of wealth, economic growth and production entail the negation of ecological integrity since the time and requirements of capitalistic production and those of the natural processes are not the same. The result is conflicts of ecological distribution, such as those found in struggles around the protection of forests, rivers, mangroves, wetlands, or biodiversity areas for protection of the livelihoods of poor communities (Escobar 2006). The protection of ecological resources is influenced by powerful factors. It refers to the process whereby nature or physical matter such as land, energy, waste, water, trees, or ocean, is transformed into useable, ownable and tradable commodities often in furtherance of capital accumulation. This socio-ecological process changes the social reproduction of urban spaces and the social relations of property in ways that increase resilience for some (powerful individuals) while expanding vulnerability and risks for others (especially the poor and marginalized) (Ajibade 2017).

This is also seen in the incorporation of environmentalism into institutions of neoliberalism which have been valuable in the rolling-out and strength owing to the fluidity and adaptability of neoliberal discourse which was evident in the Kyoto Declaration[6] and its flexible mechanisms of deploying markets (Capitalist) as the solution to environmental problems (Plastow 2010). Plastow (2010) emphasizes that the traditional approach in environmental governance has been government regulation, in what is known as 'command and control' practices, such as capping emissions or fish catches. These rely on legally established limits and are enforced by the state. However, many policies today contain elements of the neoliberal agenda and 'free market environmentalism' which has proliferated in a dialogue between environmentalism and proponents of neoliberalism, include tradeable emission permits, eco-taxes, transferable fishing quotas, user fees for public goods and aspects of utility privatization, as well as plenty of corporate green-washing (Plastow 2010). Environmental problems which are traditionally handled by the state become a commodity in the hands of the private sector.

A study by Nwokoro and Dekolo (2012) shows this nexus, that land reclamation by the Nigeria government and rich individuals especially in cities like Lagos has led to notable changes, which include the loss of forest reserves and freshwater swamps, which reduced by approximately 20% land to built-up urban development. This has also prompted the massive movement of people to the suburbs like Ikorodu and Alimosho[7] where land is cheaper and accessible (Nwokoro and Dekolo 2012).

Parker (2008) emphasis that another way of looking at the neoliberal policies is the roll-back of the state power accompanied by the roll-out of neoliberalism and the familiar near worship of the self-regulating market as the mechanism for allocating all goods and services. This dogma leads to privatization through putatively market-based schemes along with deeply contradictory endorsements of excludable, private property rights and commodification created and defended by the state. This creates environmental problems, since power structure is stripped from the government and placed in the capitalist empire (Parker 2008). The urban inequality nexus is not an irony in itself; instead, it is affected by the poor policies carried out by the government which favour some over the others.

These environmental problems unbalance the natural atmosphere which is the life supporting system for human existence, and survival provides much of the physical milieu and the raw materials required for socio-economic progress. Humanity has no choice but to interact with it. Unfortunately, human interaction, natural disaster

[6] The Kyoto Protocol is an international treaty which extends the 1992 United Nations Framework Convention on Climate Change (UNFCCC) that commits State Parties to reduce greenhouse gas emissions, based on the scientific consensus that (a) global warming is occurring and (b) it is extremely likely that human- made CO2 emissions have predominantly caused it. The Kyoto Protocol was adopted in Kyoto, Japan, on December 11, 1997 and entered into force on February 16, 2005. There are currently 192 parties to the Protocol.

[7] These are small cities in Lagos.

and climate change, are putting unprecedented pressure and impact on the quality of our environmental conditions (CBD 2016).

Environmental security is an evolving concept developed with the consciousness of the need for societal and global environmental sustainability in the face of uncontrolled environmental degradation that is witnessed in different parts of the world. Therefore, it could be looked at as an offshoot of the principle of sustainable development with the unique aim of preserving the future of our environment for the use of the present and future generations (Akiyode 2013). Ekong (2017) explained environmental sustainability as the maintenance of the factors and practices that contribute to the quality of environment on a long-term basis. It is the rates of renewable resource harvest, pollution creation and non-renewable resource depletion that can be continued indefinitely. It involves making decisions and taking actions that are in the interests of protecting the natural world, with particular emphasis on preserving the capability of the environment to support human life (Ekong 2017). The concept of sustainable development has since moved from the environmental conservationist paradigm to a holistic model which seeks to deliver basic environmental, social and economic services to all residents of a community without threatening the viability of the natural, built and social systems upon which the delivery of these services depends (Nwokoro and Dekolo 2012). Research has shown that cities and urbanization processes modify environments, often causing impaired ecology of urban riparian zones, affecting local and regional climates, and driving losses of native biodiversity and increases in non-native species (Bai et al. 2017).

According to Seitzinger et al. (2012), it is vitally important to recognize that rapid urbanization is the main driver of the Anthropocene and the inexorable increase in fossil energy use associated with its environmental problems. As the global urban population and its consumption increase, it is not only the physical use of the planet's resources, primarily from the hinterlands, that is of concern, but also the impacts on society and the environment, and these impacts occur on many scales in which many critical thresholds are crossed (Seitzinger et al. 2012). There is a significant threat of urban growth in many ways than one to the city and environment such as clearing of natural vegetation and its replacement with other activity. There has been a marked decrease in vegetation and outcrops which has been observed due to urban growth. Vegetation clearing is the major threat to biodiversity and many cause species extinctions by reducing and fragmenting available habitat (Ade and Afolabi 2013). Biodiversity plays several important roles in urban environments as per McKinney. These roles include ecosystem services such as air and water purification and amenity values such as aesthetic enjoyment and recreation. In addition, urban biodiversity has an important role in educating an increasingly urban population about nature and species conservation (McKinney 2008).

Planetary boundaries as a 'collateral concept' closely aligned or directly linked to the development of the Anthropocene, normative, and integrate aspects of the precautionary principle in the definition of a safe operating space (Brown 2017). Although the planetary boundaries are described in terms of individual quantities and separate processes, the boundaries are tightly coupled and we do not have the luxury of concentrating our efforts on any one of them in isolation from the others, for if one boundary is transgressed, then other boundaries are also under serious risk. For instance, significant land-use changes in the Amazon could influence water resources as far away as Tibet (Rockström et al. 2009). If the environment is properly managed however, it can be a productive resource to meet our socio-economic and aesthetic needs, not only for today but also for future generations. Conversely, if poorly managed, the environment could easily become hazardous and threatening to the country's survival (CBD 2016).

The urban ecosystem is human-dominated, and the changes in environmental quality and the health of urban ecosystems are largely driven and determined by concentrated human consumption of energy and natural resources (Adams et al. 2016). The environmental problems experienced in Nigeria also pose numerous consequences to the planetary boundaries framework and are very diverse. They include soil degradation, rapid deforestation, urban air and water pollution, desertification, loss of arable land, pollution of water, air, and soil from oil spills (Babanyara et al. 2010).

Rapid population growth and urbanization are of great concern to the sustainability of cities. The more people there are in a confined location, the greater the impact on the environment and pressure on resources. As such, the unwise use of the natural environment due to unawareness, poverty, and overpopulation amongst others has led to the degradation of the environment (Babanyara et al. 2010).

Spurred on by the 1987 Brundtland Report and the subsequent 1992 Rio Summit, 'sustainability' became the empty signifier that referred generically to the phantasmagorical vision of a world in which people, the economy and the environment could happily and lovingly interact in mutually supportive, cohesive, and historically reproducible manners, mediated by increasingly 'smart' technologies that would benignly micro-engineer the delicate balance between humans and nature (Swyngedouw and Kaika 2014). A sustainable city is a city designed with consideration of environmental impact, inhabited by people dedicated to minimizing required inputs of energy, water and food, and waste output of heat, air pollution like (CO^2), and water pollution. It is related to two concepts: the concept of needs and the concept of limits (Ekong 2017). Addressing human and environmental issues related to urbanization and slums is an important sustainable development issue, and various groups are working to improve the situations in slums (Givens 2015).

In achieving an inclusive urban space, effective urban planning practice can be utilized to achieve the much-anticipated inclusion, as well as reverse the trend of informality and the growth in slum development. However, it is significant to note that the transformation for better outcomes does not happen automatically, as appropriate policies that support the needs and living conditions of the urban poor must be implemented with emphasis on the importance of every individual and their community (Adeola 2016). Urban areas require policies that emphasize not only economic growth but also balancing sustainability along three domains; (people, profit, and the planet) (Jack et al. 2017). Land use and land cover changes due to city expansion have serious ecological repercussions and pose significant challenges to environmental sustainability at local and global levels. The last few decades have witnessed a threat potentially undermining the capacity of ecosystems to sustain food production, maintain freshwater and forest resources, regulate the climate, and air quality. We are faced with the challenge of ensuring "sustainable development", which tends to manage the trade-offs between immediate human needs and maintaining the capacity of the biosphere to provide goods and services in the long term (Nwokoro and Dekolo 2012).

Conclusion

Introduction

Sub-Saharan Africa finds itself at a critical turning point, confronting the intersecting challenges of rapid urbanization, climate change, and the pressing need for sustainable development. This book has explored the diverse yet interconnected themes of urban green infrastructure, renewable energy, and sustainable urbanization, all of which are crucial to the region's future. By examining these issues in varied contexts—whether in the informal settlements of expanding cities, the energy landscapes of Cameroon, or the urban development challenges in Nigeria—this exploration has highlighted both the significant obstacles and the promising opportunities that lie ahead on Sub-Saharan Africa's path toward sustainability.

In the first part, we delved into the essential role of urban green infrastructure, particularly in the context of Sub-Saharan Africa's informal settlements, which are especially vulnerable to the intensifying effects of climate change, such as extreme heat. These settlements, often marked by dense populations and limited infrastructure, urgently need sustainable solutions. Urban green infrastructure—whether through parks, tree-lined streets, or gardens—goes beyond mere beautification; it is a crucial tool for regulating urban temperatures and improving the quality of life for city dwellers. The physical effects of extreme heat extend to public health, energy use, and overall social well-being. The case studies in this section demonstrated that green infrastructure could help mitigate these challenges but ensuring that these solutions benefit all members of society, particularly the marginalized, remains a significant hurdle. Limited funding, maintenance concerns, and the need for community involvement are some of the persistent challenges to be addressed.

In the second part, we turned to renewable energy, using Cameroon as a focal point to explore broader regional implications. Renewable energy sources such as solar, wind, hydro, and biomass offer substantial opportunities for Sub-Saharan Africa to enhance its energy security and foster economic growth while reducing its environmental footprint. Cameroon's journey toward renewable energy reveals both the

© The Editor(s) (if applicable) and The Author(s), under exclusive license to Springer Nature Singapore Pte Ltd. 2025

R. Perrault et al., *Sub-Saharan African Sustainable Development*, City Development: Issues and Best Practices, https://doi.org/10.1007/978-981-95-1458-8

challenges and the potential of this transition. The country's energy landscape is shaped by governance, policy frameworks, and infrastructure, all of which present obstacles such as regulatory constraints, financial limitations, and technical capacity. However, the promise of renewable energy in promoting sustainability, enhancing resilience, and improving livelihoods is significant. Moreover, the socio-ecological impacts of renewable projects—such as the effect of hydropower on local ecosystems and communities—underscore the need for careful planning that integrates social and environmental considerations.

The final part focused on Nigeria, a nation emblematic of the rapid urbanization trends across Sub-Saharan Africa. The historical forces shaping Nigeria's urban development—from colonial legacies to demographic pressures—have resulted in sprawling cities grappling with sustainability issues. Deforestation, pollution, overcrowding, and inadequate infrastructure are among the urgent challenges facing urban areas. Social issues such as unemployment, insufficient housing, and the rise of informal settlements further complicate the urban landscape. Yet, despite these daunting challenges, there are also signs of progress. Various sustainable development initiatives—from waste management programs to urban agriculture and renewable energy projects—highlight the potential for both top-down and community-driven solutions. The lessons learned from these efforts emphasize the need for scalability, adaptability, and continuous community engagement.

Overarching Themes and Imperatives for Sustainable Development

Throughout this exploration, several key themes emerged, providing a blueprint for sustainable development in Sub-Saharan Africa. First, the need for holistic and integrated approaches is clear. Isolated initiatives will not suffice; strategies must address environmental conservation, economic growth, and social well-being simultaneously. For example, urban green infrastructure does more than improve environmental outcomes—it also fosters social cohesion and economic opportunities.

Equally important is the role of community engagement and inclusivity. Sustainable solutions must be rooted in the needs and realities of local communities, with participatory planning processes that respect indigenous knowledge and ensure that all stakeholders, including marginalized groups like women and the urban poor, benefit from development efforts. Empowering communities to shape their own development is vital for ensuring equitable outcomes. Governance also plays a critical role. Adaptive and responsive governance structures are essential for sustainable development to take root. Governments must be transparent, accountable, and willing to collaborate with the private sector, civil society, and international partners. This cooperation strengthens the capacity to address the complex and intertwined challenges of sustainability. Knowledge sharing and capacity building further contribute to the longevity of development initiatives. Investing in education and training ensures that best practices are disseminated across the region and that communities are empowered to take ownership of projects. Resilience and adaptability, particularly in the face of climate change, are likewise critical. As the region faces increasing environmental uncertainties, flexible strategies will be needed to absorb shocks and minimize disruptions to livelihoods and ecosystems.

The Road Ahead: Opportunities and Challenges

While the challenges facing Sub-Saharan Africa are immense, the opportunities for innovation and transformation are equally significant. The region's youthful population, abundant natural resources, and emerging technological advancements present a unique moment to redefine development trajectories. Technological innovation can be a powerful accelerator of sustainable development, with advancements in renewable energy, smart urban planning, and sustainable agriculture offering new pathways for progress. Economic diversification—moving away from extractive industries toward sectors like renewable energy, ecotourism, and sustainable agriculture—will also be critical to fostering resilience and inclusive growth.

Regional integration presents another important opportunity. Strengthening cooperation through shared infrastructure, coordinated policies, and integrated markets can amplify the benefits of sustainable development initiatives. International partnerships, too, have a crucial role to play, offering access to financial resources, technical expertise, and best practices. However, these partnerships must be grounded in mutual respect and aligned with locally defined priorities.

To harness these opportunities and navigate the challenges ahead, several recommendations stand out. First, policy frameworks that promote sustainable practices across sectors need to be strengthened. Investments in sustainable infrastructure, such as renewable energy systems and green urban spaces, should be prioritized. Education and public awareness are also essential, empowering communities with the knowledge and skills they need to participate actively in development processes. Strengthening institutions and building capacity are equally important, ensuring that those tasked with implementing sustainable initiatives are equipped to succeed.

Monitoring and evaluation systems must be robust, ensuring accountability and facilitating continuous learning. Inclusive economic policies are necessary to promote equity and access, particularly for marginalized groups. Finally, climate change mitigation and adaptation strategies must be comprehensive and community-led development must be supported, as local initiatives are often more sustainable and effective.

A Vision for Sustainable Sub-Saharan Africa

The journey toward sustainable development in Sub-Saharan Africa is both intricate and transformative. It demands a collective commitment to rethink development strategies that harmonize economic growth, environmental protection, and social equity. Throughout this book, we have explored the critical roles of urban green infrastructure, renewable energy development, and sustainable urbanization. These elements are not merely areas of policy focus but pivotal components of a broader vision for a resilient and sustainable future in the region.

Sub-Saharan Africa holds immense potential to become a global leader in sustainable development, drawing on its rich natural resources, youthful population, and capacity for innovation. However, realizing this vision will require integrated, inclusive approaches, technological advancements, and strengthened governance. The path forward is not without its challenges—ranging from financial limitations to regulatory hurdles—but with collaborative effort and a focus on long-term goals, a sustainable and prosperous Sub-Saharan Africa is attainable.

In this envisioned future, cities are vibrant, livable, and resilient. Energy systems are clean, accessible, and equitable, ensuring that development benefits all sectors of society. Achieving such a vision demands more than policy reforms or infrastructure investments. It requires a shift in collective thinking, prioritizing sustainability over short-term gains. Policymakers, practitioners, scholars, and communities must work together, continuously adapting and learning as they navigate the region's complex challenges.

Sustainable development is an evolving process that must adapt to changing circumstances and new insights. As we conclude this exploration, it is important to recognize that the ideas and recommendations in this book are intended to inspire ongoing dialogue and spur action across Sub-Saharan Africa and beyond.

The Call to Action

To turn this vision into reality, concerted action from various stakeholders is essential. Governments must prioritize sustainable development by embedding it into national agendas, allocating necessary resources, and fostering environments where innovation and public participation thrive. The private sector has a key role in adopting sustainable practices, investing in green technologies, and engaging in corporate social responsibility efforts that align with broader sustainability goals.

Civil society organizations are instrumental in advocating for sustainability, raising awareness, and ensuring that governments and businesses are held accountable to their commitments. Academic and research institutions should continue to generate knowledge, inform policymaking, and provide evidence-based solutions to the sustainability challenges that the region faces. At the community level, individuals and local organizations must be empowered to take an active role in development processes, adopting sustainable practices and contributing to the building of resilient societies.

Final Thoughts

The challenges facing Sub-Saharan Africa in its pursuit of sustainable development are significant, but they are by no means insurmountable. With a shared vision and collaborative effort, it is entirely possible to create a future in which economic prosperity, social well-being, and environmental health are not only aligned but mutually reinforcing. This book serves as both a reflection on the current state of affairs in the region and a roadmap for the journey ahead.

Let this be a call for sustained dialogue, continuous innovation, and decisive action. A sustainable Sub-Saharan Africa is within reach, but it will require determination, creativity, and collective effort. By working together—across governments, private sectors, civil society, academia, and communities—the promise of a sustainable and equitable future for all can be fulfilled. The journey is ongoing, but with commitment, adaptability, and perseverance, the vision of a thriving and sustainable Sub-Saharan Africa can become a reality.

References

Abanda FH (2012) Renewable energy sources in Cameroon: potentials, benefits and enabling environment. Renew Sustain Energy Rev 16(7):4557–4562. https://doi.org/10.1016/j.rser.2012.04.011

Abbass IM (2012) Trends of rural-urban migration in Nigeria. Eur Sci J 8(3):97–125

Adams S, Adom PK, Klobodu EK (2016) Urbanization, regime type and durability, and environmental degradation in Ghana. Environ Sci Pollut Res Int 23(23):23825–23839. https://doi.org/10.1007/S11356-016-7513-4

Ade MA, Afolabi YD (2013) Monitoring urban sprawl in the Federal Capital Territory of Nigeria using remote sensing and GIS techniques. Ethiop J Environ Stud Manag 6(1):82–95

Adebayo C (2017) When climate change hit Makoko—the Lagos slum. Daily Post Nigeria. http://Dailypost.Ng/2017/09/29/Caleb-Adebayo-Climate-Change-Hit-Makoko-Lagos-Slum/. Retrieved 8 Jan 2018

Adegeye AJ (1982) Establishing river basin development authorities as a strategy for Nigerian rural development. Agric Adm 9(4):301–311. https://doi.org/10.1016/0309-586x(82)90048-6

Adegun OB (2019) Green infrastructure in informal unplanned settlements: the case of Kya Sands, Johannesburg. Int J Urban Sustain Dev 11(1):68–80. https://doi.org/10.1080/19463138.2019.1565412

Adegun OB, Mbuya EC, Njavike E (2022) Responses to heat stress within an unplanned settlement in Dar Es Salaam, Tanzania. Front Built Environ. https://doi.org/10.3389/fbuil.2022.874751

Adelekan I (2011) Vulnerability assessment of an urban flood in Nigeria: Abeokuta Flood 2007. Nat Hazards 56(1):215–231. https://doi.org/10.1007/S11069-010-9564-Z

Adeloye AJ, Rustum R (2011) Lagos (Nigeria) flooding and influence of urban planning. Proc Ice Urban Des Plan 164(3):175–187. https://doi.org/10.1680/Udap.1000014

Adeola OS (2016) Urban egalitarianism: the way forward to ensure sustainable urban design, practice and development in developing countries (The Nigeria Case). Environ Ecol Res 4(3):107–115. https://doi.org/10.13189/Eer.2016.040301

Adger WN (2006) Vulnerability. Glob Environ Change 16(3):268–281. https://doi.org/10.1016/j.gloenvcha.2006.02.006

Adom PK (2016) The transition between energy efficient and energy inefficient states in Cameroon. Energy Econ 54:248–262. https://doi.org/10.1016/j.eneco.2015.11.025

Aduwo BE, Edewor P, Ibem EO (2016) Urbanization and housing for low-income earners in Nigeria: a review of features, challenges and prospects. Mediterr J Soc Sci 7(3):347–357

African Union (2010) Cameroon axes illegal logging | The Africa-EU Partnership. https://www.africa-eu-partnership.org/en/stay-informed/news/cameroon-axes-illegal-logging. Retrieved 25 March 2019

© The Editor(s) (if applicable) and The Author(s), under exclusive license to Springer Nature Singapore Pte Ltd. 2025

R. Perrault et al., *Sub-Saharan African Sustainable Development*, City Development: Issues and Best Practices, https://doi.org/10.1007/978-981-95-1458-8

African Union Commission (2022) African Union climate change and resilient development strategy and action plan (2022–2032). African Union, p 128. https://au.int/en/documents/20220628/african-union-climate-change-and-resilient-development-strategy-and-action-plan
Agbola BS, Ajayi O, Taiwo OJ, Wahab BW (2012) The August 2011 flood in Ibadan, Nigeria: anthropogenic causes and consequences. Int J Disaster Risk Sci 3(4):207–217
Agboola AM, Ayanlade A (2016) Sea level rise and its potential impacts on coastal urban area: a case of Eti-Osa, Nigeria. Ann Univ Oradea Geogr Ser/Analele Universitatii Din Oradea, Seria Geografie 26(2):188–200
Agbu O (2007) The iron and steel industry and Nigeria's industrialization: exploring cooperation with japan. Institute of Developing Economies, Japan External Trade Organization. Retrieved from http://Www.Worldcat.Org/Title/Iron-and-Steel-Industry-and-Nigerias-Industrialization-Exploring-Cooperation-with-Japan/Oclc/181518994
Ajayi OO, Oviasogie FO, Azuh DE, Duruji MM (2014) Urban design and sustainable development: a case of Makoko area of Lagos State, Nigeria. Eur Sci J 2:90–97
Ajibade I (2017) Can a future city enhance urban resilience and sustainability? A political ecology analysis of Eko Atlantic City, Nigeria. Int J Disaster Risk Reduct 26:85–92. https://doi.org/10.1016/J.Ijdrr.2017.09.029
Ajibola MO, Adewale BA, Ijasan KC (2012) Effects of urbanisation on Lagos wetlands. Int J Bus Soc Sci 3(17):310–318
Akanle O, Adejare GS (2017) Conceptualising megacities and mega slums in Lagos, Nigeria. Africa's Public Serv Deliv Perform Rev 5(1):1–9. https://doi.org/10.4102/Apsdpr.V5i1.155
Akinsete E, Hoelzel F, Oshodi L (2014) Delivering sustainable urban regeneration in emerging nations: introducing neighborhood hotspots. J Archit Educ 68(2):238–245. https://doi.org/10.1080/10464883.2014.937245
Akiyode OO (2013) Implications of urbanization on environmental security in developing economy countries: a case study of Nigeria. J Sustain Dev Africa 15(3):1–12
Ali H, Law S, Zannah T (2016) Dynamic impact of urbanization, economic growth, energy consumption, and trade openness on CO2 emissions in Nigeria. Environ Sci Pollut Res 23(12):12435–12443. https://doi.org/10.1007/S11356-016-6437-3
Amao FL, Ilesanmi AO (2013) Housing quality in the urban fringes of Ibadan, Nigeria. In: Sustainable building conference, pp 149–158. https://www.irbnet.de/daten/iconda/CIB_DC26503.pdf
Amigun B, Musango JK, Brent AC (2011) Community perspectives on the introduction of biodiesel production in the Eastern Cape Province of South Africa. Energy 36(5):2502–2508. https://doi.org/10.1016/j.energy.2011.01.042
Antonel J, Chowdhury Z (2014) Measuring ambient particulate matter in three cities in Cameroon, Africa. Atmos Environ 95:344–354
ARSEL (2018) La transparence pour un secteur électrique performant—Transparency for a performant electricity sector. (n.d.). http://arsel-cm.org/?lang=en. Retrieved 8 Nov 2018
Asaju K (2017) Sustainable rural development in Nigeria: an overview. Online J Arts Manag Soc Sci 2(1):173–182
Ashukem J-CN (2016) Included or excluded: an analysis of the application of the free, prior and informed consent principle in land grabbing cases in Cameroon. Potchefstroom Electron Law J/Potchefstroomse Elektroniese Regsblad 19(1)
Asumadu-Sarkodie S, Owusu PA (2016) Forecasting Nigeria's energy use by 2030, an econometric approach. Energy Sources Part B 11(10):990–997. https://doi.org/10.1080/15567249.2016.1217287
Ayedun CA, Durodola OD, Akinjare OA (2011) Towards Ensuring sustainable urban growth and development in Nigeria: challenges and strategies. Bus Manag Dyn 1(2):99–104
Ayeni AO (2014) Domestic water source, sanitation and high risk of bacteriological diseases in the urban slum: case of Cholera in Makoko, Lagos, Nigeria. J Environ Pollut Hum Health 2(1):12–15

Ayompe LM, Duffy A (2014) An assessment of the energy generation potential of photovoltaic systems in Cameroon using satellite-derived solar radiation datasets. Sustain Energy Technol Assess 7:257–264. https://doi.org/10.1016/j.seta.2013.10.002

Azongo DK, Awine T, Wak G, Binka FN, Rexford Oduro A (2012) A time series analysis of weather variables and all-cause mortality in the Kasena-Nankana Districts of Northern Ghana, 1995–2010. Glob Health Action 5(1):19073. https://doi.org/10.3402/gha.v5i0.19073

Babalobi B (2013) Water, sanitation and hygiene practices among primary-school children in Lagos: a case study of the Makoko Slum Community. Water Int 38(7):921–929. https://doi.org/10.1080/02508060.2013.851368

Babanyara YY, Usman HA, Saleh UF (2010) An overview of urban poverty and environmental problems in Nigeria. J Hum Ecol 31(2):135–143

Badgley C (2009) Cameroon: pipeline to prosperity? http://www.pbs.org/frontlineworld/stories/bribe/2010/06/ten-years-ago-this-month.html. Retrieved 25 March 2019

Bai X, Mcphearson T, Cleugh H, Nagendra H, Tong X, Zhu T, Zhu Y-G (2017) Linking urbanization and the environment: conceptual and empirical advances. Annu Rev Environ Resour 42:215–240. https://doi.org/10.1146/Annurev-Environ-102016-061128

Baird I, Flaherty M, Baird I (2005) Mekong river fish conservation zones in Southern Laos: assessing effectiveness using local ecological knowledge. Environ Manag 36(3):439–454. https://doi.org/10.1007/s00267-005-3093-7

Bele MY, Somorin O, Sonwa DJ, Nkem JN, Locatelli B (2011) Forests and climate change adaptation policies in Cameroon. Mitig Adapt Strat Glob Change 16(3):369–385. https://doi.org/10.1007/s11027-010-9264-8

Bele MY, Sonwa DJ, Tiani AM (2013) Supporting local adaptive capacity to climate change in the Congo basin forest of Cameroon. Int J Clim Change Strateg Manag 5(2):181–197. https://doi.org/10.1108/17568691311327587

Berkes F (2009) Indigenous ways of knowing and the study of environmental change. J R Soc N Z 39(4):151–156. https://doi.org/10.1080/03014220909510568

Berkes F, Colding J, Folke C (2000) Rediscovery of traditional ecological knowledge as adaptive management. Ecol Appl 10(5):1251–1262. https://doi.org/10.1890/1051-0761(2000)010[1251:ROTEKA]2.0.CO;2

Blaikie P, Brown K, Stocking M, Tang L, Dixon P, Sillitoe P (1997) Knowledge in action: local knowledge as a development resource and barriers to its incorporation in natural resource research and development. Agric Syst 55(2):217–237. https://doi.org/10.1016/S0308-521X(97)00008-5

Bloch R, Monroy J, Fox S, Ojo A (2015) Urbanisation and urban expansion in Nigeria. Urbanisation Research Nigeria (URN) Research Report. London: ICF International. Creative Commons Attribution-NonCommercial-ShareAlike CC BY-NC-SA. http://urn.icfwebservices.com/Media/Default/Publications/URN%20Theme%20A%20%Urbanisation%20Report%20FINAL.pdf

Brown K (2017) Global environmental change II: planetary boundaries—a safe operating space for human geographers? Prog Hum Geogr 41(1):118–130. https://doi.org/10.1177/0309132515604429

Bukola Bakare B, Ige T (2017) Flood wreaks havoc in Nigeria's most expensive neighbourhoods. Punch. https://punchng.com/flood-wreaks-havoc-in-nigerias-most-expensive-neighbourhoods/. Retrieved 17 Sept 2017

Calzadilla P, Mauger R (2018) The UN's new sustainable development agenda and renewable energy: the challenge to reach SDG7 while achieving energy justice. J Energy Nat Resour Law 36(2):233–254. https://doi.org/10.1080/02646811.2017.1377951

Clearing House Mechanism of Nigeria (2016) Main threats to biodiversity in Nigeria. *Convention on Biological Diversity (CBD).* http://www.Chm-Cbd.Com.Ng/Index.Php/En/Biodiversity/Main-Threats-To-Biodiversity-In-Nigeria/52-Other-Relevant-Resources. Retrieved 8 Jan 2018

Carmody P, Owusu FY (2016) Neoliberalism, urbanization and change in Africa: the political economy of heterotopias. J Afr Dev 18(18):61–73

Catney P, Dobson A, Hall SM, Hards S, MacGregor S, Robinson Z et al (2013) Community knowledge networks: an action-orientated approach to energy research. Local Environ 18(4):506–520. https://doi.org/10.1080/13549839.2012.748729

Cavan G, Lindley S, Jalayer F, Yeshitela K, Pauleit S, Renner F, Gill S, Capuano P, Nebebe A, Woldegerima T, Kibassa D, Shemdoe R (2014) Urban morphological determinants of temperature regulating ecosystem services in two African cities. Ecol Ind 42:43–57. https://doi.org/10.1016/j.ecolind.2014.01.025

Chen M, Zhang H, Liu W, Zhang W (2014) The global pattern of urbanization and economic growth: evidence from the last three decades. PLoS ONE 9(8):1–15. https://doi.org/10.1371/journal.pone.0103799

Chothia F (2018) Red dragons and tigers—Cameroon's English-speaking rebels. Retrieved from https://www.bbc.com/news/world-africa-45723211

Cities in Nigeria, Nigeria Map with Cities (2017) Mapofworld. https://www.Mapsofworld.Com/Nigeria/Cities/. Retrieved 15 March 2018

Coaffee J (2008) Risk, resilience, and environmentally sustainable cities. Energy Policy 36(12):4633–4638. https://doi.org/10.1016/j.enpol.2008.09.048

Colucci AM (2016) Peri-urban areas and food-energy-water nexus: sustainability and resilience strategies in the age of climate change. In: Springer tracts in civil engineering. Retrieved from https://doi.org/10.1007/978-3-319-41022-7

Comaroff J (2014) Built on sand: Singapore and the new state of risk. Harvard Design Magazine. http://www.Harvarddesignmagazine.Org/Issues/39/Built-On-Sand-Singapore-And-The-New-State-Of-Risk. Retrieved 14 Oct 2017

Comeau LA, Parkins JR, Stedman RC, Beckley TM (2015) Citizen perspectives on energy issues in Canada: a national survey of energy literacy and energy citizenship. Retrieved from http://ageconsearch.umn.edu/bitstream/211095/2/PR-15-01.pdf

Cook CC, Bank AD (2005) Assessing the impact of transport and energy infrastructure on poverty reduction. Asian Development Bank

Corburn J (2003) Bringing local knowledge into environmental decision making: improving urban planning for communities at risk. J Plan Educ Res 22(4):420–433

Country Profiles (2017) Global forest watch. https://www.globalforestwatch.org/dashboards/country/NGA. Retrieved 8 Jan 2018

Creighton S (2014) Slum on stilts. Daily Mail. https://www.dailymail.co.uk/news/article-2608506/Slum-stilts-A-mother-baby-paddles-dirty-oily-water-Nigerias-Makoko-slum-250-000-residents-huddle-homes-lagoon.html. Retrieved Sept 2017

Cyril S, Oldroyd JC, Renzaho A (2013) Urbanisation, urbanicity, and health: a systematic review of the reliability and validity of urbanicity scales. BMC Public Health 13(513):1–11. https://doi.org/10.1186/1471-2458-13-513

Devine-Wright P (2009) Rethinking NIMBYism: the role of place attachment and place identity in explaining place-protective action. J Community Appl Soc Psychol 19(6):426–441. https://doi.org/10.1002/casp.1004

Di Leo N, Escobedo FJ, Dubbeling M (2016) The role of urban green infrastructure in mitigating land surface temperature in Bobo-Dioulasso, Burkina Faso. Environ Dev Sustain 18(2):373–392. https://doi.org/10.1007/s10668-015-9653-y

Djiofack CZ, Omgba LD (2011) Oil depletion and development in Cameroon: a critical appraisal of the permanent income hypothesis. Energy Policy 39(11):7202–7216. https://doi.org/10.1016/j.enpol.2011.08.041

Dominguez-Torres C, Foster V (2011) Cameroon's infrastructure: a continental perspective 59

Drescher M (2019) Urban heating and canopy cover need to be considered as matters of environmental justice. Proc Natl Acad Sci 116(52):26153–26154. https://doi.org/10.1073/pnas.1917213116

du Toit MJ, Cilliers SS, Dallimer M, Goddard M, Guenat S, Cornelius SF (2018) Urban green infrastructure and ecosystem services in sub-Saharan Africa. Landsc Urban Plan 180:249–261. https://doi.org/10.1016/j.landurbplan.2018.06.001

Ebong C, Levy P (2011) Impact of the introduction of new vaccines and vaccine wastage rate on the cost-effectiveness of routine EPI: Lessons from a descriptive study in a Cameroonian health district, vol 9. https://doi.org/10.1186/1478-7547-9-9
Ejaro SP, Abubakar A (2013) The challenges of rapid urbanization on sustainable development of Nyanya, Federal Capital Territory, Abuja, Nigeria. J Appl Sci Environ Manag 17(2):299–313
Ekong FU (2017) Applying the concept of green cities in Nigeria: challenges and prospects. Adv Soc Sci Res J 4(10):85–96
Elisha Sawe B (2017) Biggest cities in Nigeria. World Atlas. https://www.worldatlas.com/articles/biggest-cities-in-nigeria.html. Retrieved Sept 2017
Emery FE (1975) Towards a social ecology: contextual appreciations of the future in the present. Plenum PubCorp, New York
Enerray (2018) Le photovoltaïque devient service public au Cameroun. https://www.enerray.com/fr/blog/photovoltaique-en-cameroun. Retrieved 22 Nov 2018
Eni DI, Obiefuna JN, Oko C, Ekwok I (2011) Impact of urbanization on subsurface water quality in Calabar Municipality, Nigeria. Int J Human Soc Sci 1(10):167–172
Enongene T, Moses J et al (2016) Enhancing urban energy security through renewable energy solutions: case study Douala, Cameroon. Retrieved from https://www.theseus.fi/handle/10024/121725
Epule TE, Ford JD, Lwasa S, Lepage L (2017) Climate change adaptation in the Sahel. Environ Sci Policy 75:121–137. https://doi.org/10.1016/j.envsci.2017.05.018
Escobar A (2006) Difference and conflict in the struggle over natural resources: a political ecology framework. Development 49(3):6–13. https://doi.org/10.1057/palgrave.development.1100267
ESMAP (2007) Energy sector management assistance program technical paper. Plan d'Action National Enerhie pour la Réduction de la Pauvreté 117/07 FR, Cameroun
Etonde (2018) Cameroon: local women besieged by the military, guardians of the palm oil plantations of SOCAPALM | WRM in English. https://wrm.org.uy/articles-from-the-wrm-bulletin/section1/cameroon-local-women-besieged-by-the-military-guardians-of-the-palm-oil-plantations-of-socapalm/. Retrieved 19 Jan 2019
FAO (2018) SOFI 2018—the state of food security and nutrition in the World. www.fao.org website. http://www.fao.org/state-of-food-security-nutrition/en/. Retrieved 2 Apr 2019
FAO Country Profiles: Nigeria (2017) Food and agriculture organization of the United Nations. http://www.Fao.Org/Countryprofiles/Index/En/?Iso3=Nga. Retrieved 8 Jan 2018
Farfan J, Peltoniemi P, Breyer C (2018) Repercussion of large scale Hydro Dam deployment: the case of Congo Grand Inga Hydro Project. Energies 11(4):1–30. https://doi.org/10.3390/en11040972
Feyisa GL, Dons K, Meilby H (2014) Efficiency of parks in mitigating urban heat island effect: an example from Addis Ababa. Landsc Urban Plan 123:87–95. https://doi.org/10.1016/j.landurbplan.2013.12.008
Fondja Wandji YD (2013) Energy consumption and economic growth: evidence from Cameroon. Energy Policy 61:1295–1304. https://doi.org/10.1016/j.enpol.2013.05.115
Fonjong L, Emmanuel N, Fonchingong C (2005) Rethinking the contribution of indigenous management in small-scale water provision among selected rural communities in Cameroon. Environ Dev Sustain 6(4):429–451. https://doi.org/10.1007/s10668-005-0501-3
Fonjong L, Sama-Lang I, Fombe L, Abonge C (2016) Land governance and women's rights in large-scale land acquisitions in Cameroon. Dev Pract 26(4):420–430. https://doi.org/10.1080/09614524.2016.1162285
Forest Data: Nigeria Deforestation Rates and Related Forestry Figures (2006) Mongabay. https://Rainforests.Mongabay.Com/Deforestation/Archive/Nigeria.Htm. Retrieved 8 Jan 2018
Forsyth T (2003) Critical political ecology: the politics of environmental science. Routledge, New York. https://doi.org/10.4324/9780203017562
Fox S (2017) Mortality, migration, and rural transformation in Sub-Saharan Africa's urban transition. J Demogr Econ Louvain 83(1):13–30. https://doi.org/10.1017/Dem.2016.29

Funwie (2018) Une nouvelle centrale solaire photovoltaïque mise en service au Cameroun—Médiaterre (n.d.). https://www.mediaterre.org/afrique-centrale/actu,20180827122131.html. Retrieved 22 Nov 2018
Gabriel F, Abraham T (2011) Urbanization, housing and infrastructural facilities in Lagos, Nigeria. Dimensi (J Archi Built Environ) 37(1):9–14
Garschagen M, Romero-Lankao P (2015) Exploring the relationships between urbanization trends and climate change vulnerability. Clim Change 133(1):37–52. https://doi.org/10.1007/S10584-013-0812-6
Gaudreau M, Cao H (2015) Political constraints on adaptive governance: environmental NGO networks in Nanjing, China. J Environ Dev 24(4):418–444. https://doi.org/10.1177/1070496515602044
Gaye A (2007) OCCASIONAL PAPER Access to energy and human development
Gbadegesin JT, Oladokun TT, Ayorinde OI (2011) Urban renewal as a tool for sustainable urban development in Nigeria: issues and challenges. J Sustain Dev Environ Prot 1(1):57–68
Gbetnkom D (2005) Deforestation in Cameroon: immediate causes and consequences. Environ Dev Econ 10(4):557–572. https://doi.org/10.1017/S1355770X05002330
Giordano R, Liersch S, Vurro M, Hirsch D (2010) Integrating local and technical knowledge to support soil salinity monitoring in the Amudarya river basin. J Environ Manag 91(8):1718–1729. https://doi.org/10.1016/j.jenvman.2010.03.010
Givens JE (2015) Urbanization, slums, and the carbon intensity of well-being: implications for sustainable development. Hum Ecol Rev 22(1):107–128
Global Center on Adaptation (2022) State and trends in adaptation report 2022. Global Center on Adaptation. https://gca.org/reports/sta22/
Global Forest Watch (2017). https://www.globalforestwatch.org/dashboards/country/CMR?category=summary&treeLossTsc=eyJoaWdobGlnaHRlZCI6ZmFsc2V9. Retrieved 3 April 2019
Goldthau A (2014) Rethinking the governance of energy infrastructure: Scale, decentralization and polycentrism. Energy Res Soc Sci 1:134–140. https://doi.org/10.1016/j.erss.2014.02.009
Green Peace (2015) Herakles Farms project rears its ugly head again (n.d.). Greenpeace International website. https://www.greenpeace.org/international/story/6959/herakles-farms-project-rears-its-ugly-head-again. Retrieved 2 Apr 2019
Greenpeace International Congo Basin (2017) http://www.greenpeace.org/international/en/campaigns/forests/africa/. Retrieved 16 July 2017
Grijsen J (2014) Understanding the Impact of Climate Change on Hydropower: the case of Cameroon, 166
GTNM web site (2018) https://www.mng-cameroon.org/index.html. Retrieved 8 Nov 2018
Haggett C (2011) Understanding public responses to offshore wind power. Energy Policy 39(2):503–510. https://doi.org/10.1016/j.enpol.2010.10.014
Hassan MM, Kalam MA, Shano S, Nayem MRK, Rahman MK, Khan SA, Islam A (2020) Assessment of epidemiological determinants of COVID-19 pandemic related to social and economic factors globally. J Risk Financ Manag 13(9), Article 9. https://doi.org/10.3390/jrfm13090194
Heynen N (2014) Urban political ecology I: the urban century. Prog Hum Geogr 38(4):598–604. https://doi.org/10.1177/0309132513500443
Heynen N, Kaika M, Swyngedouw E (2005) Urban political ecology: politicizing the production of urban natures. In: In the nature of cities: urban political ecology and the politics of urban metabolism. Taylor & Francis Group, pp 1–20. http://ebookcentral.proquest.com/lib/ottawa/detail.action?docID=256935
Heynen N, Kaika M, Swyngedouw E (2006) In the nature of cities urban political ecology and the politics of urban metabolism. Taylor and Francis. Print, New York, NY
Hiemstra-van der Horst G, Hovorka AJ (2008) Reassessing the "energy ladder": household energy use in Maun, Botswana. Energy Policy 36(9):3333–3344. https://doi.org/10.1016/j.enpol.2008.05.006
Hodbod J, Adger WN (2014) Integrating social-ecological dynamics and resilience into energy systems research. Energy Res Soc Sci 1:226–231. https://doi.org/10.1016/j.erss.2014.03.001

Huang S-L, Chen C-W (2005) Theory of urban energetics and mechanisms of urban development. Ecol Model 189(1–2):49–71. https://doi.org/10.1016/j.ecolmodel.2005.03.004

Hunter NB, North MA, Roberts DC, Slotow R (2020) A systematic map of responses to climate impacts in urban Africa. Environ Res Lett 15(10):103005. https://doi.org/10.1088/1748-9326/ab9d00

Huud S (2015) Trend of Islamicology in Nigeria; past, present and future. A paper for 2nd International conference on Arabic studies and Islamic civilization between 9th and 10th March organised by Department of Arabic Studies and Linguistic, Academy of Islam, International Islamic University College. March 9th and 10th, 2015. Kuala Lumpor, Malaysia. https://worldconferences.net/proceedings/icasic2015/full%20paper/IC%20323%20TREND%20OF%20ISLAMICOLOGY%20IN%20NIGERIA%202.pdf

Ibem EO (2011) Challenges of disaster vulnerability reduction in Lagos Megacity Area, Nigeria. Disaster Prev Manag Int J 20(1):27–40. https://doi.org/10.1108/09653561111111063

Idowu TE, Home P (2017) Probable effects of sea level rise and land reclamation activities on coastlines and wetlands of Lagos Nigeria. In: The 2015 JKUAT scientific conference, pp 207–220. http://journals.jkuat.ac.ke/index.php/jscp/article/viewFile/1251/1028

Imam M, Jamasb T, Llorca M, Llorca M (2018) Power sector reform and corruption: evidence from electricity industry in sub-Saharan Africa. IDEAS Working Paper Series from RePEc. Retrieved from http://search.proquest.com/docview/2059019784/?pq-origsite=primo

Innocent EO (2013) Urban planning problems in Nigeria: a case of Onitsha Metopolis of Anambra State. Singap J Bus Econ Manag Stud 1(1):41–56

International Conference on Canadian, Chinese and African Sustainable Urbanization (2017) ICCUSU II; Smart urban development: from local to global actions. Yaoundé, Cameroon, 12–14 Dec 2017

International Crisis Group (2018) Cameroon's far north: a new chapter in the fight against Boko Haram (2018). https://www.crisisgroup.org/africa/central-africa/cameroon/263-extreme-nord-du-cameroun-nouveau-chapitre-dans-la-lutte-contre-boko-haram. Retrieved 28 Feb 2019

International Energy Association (2017). https://www.iea.org/access2017/. Retrieved 25 Oct 2018

International Energy Association: Statistics | Cameroon—Total Primary Energy Supply (TPES) by source (chart) (2016) https://www.iea.org/statistics/?country=CAMEROON&year=2016. Retrieved 8 Jan 2019

International Energy Association: Statistics | Cameroon—Total final consumption (TFC) by source (chart) (2016). https://www.iea.org/statistics/?country=CAMEROON&year=2016. Retrieved 22 Jan 2019

International Rivers (2005) Drought could cripple Cameroon's hydro–heavy energy sector. https://www.internationalrivers.org/resources/drought-could-cripple-cameroon%E2%80%99s-hydro%E2%80%93heavy-energy-sector-2663. Retrieved 3 Dec 2018

International Rivers (2007) Drought could cripple Cameroon's hydro–heavy energy sector. (n.d.). International Rivers website. https://www.internationalrivers.org/resources/drought-could-cripple-cameroon%E2%80%99s-hydro%E2%80%93heavy-energy-sector-2663. Retrieved 3 Dec 2018

IPCC (2007) Climate change 2007: the physical science basis: contribution of working group I to the fourth assessment report of the intergovernmental panel on climate change. Cambridge University Press, p 976. https://www.ipcc.ch/site/assets/uploads/2018/03/ar4_wg2_full_report.pdf

Isaksson C, Sumasgutner P (2016) How rapid urbanisation is changing the profile of wildlife in Cities. The Conversation. https://Theconversation.Com/How-Rapid-Urbanisation-Is-Changing-The-Profile-Of-Wildlife-In-Cities-58818. Retrieved 8 Jan 2018

Isma'il M, Ishaku E, Yahaya AM, Tanko MA, Ahmed HT (2015) Urban growth and housing problems in Karu Local Government Area of Nasarawa State. Nigeria. Glob J Res Rev 2(1):45–57

Jack M, Coles AM, Piterou A (2017) Sustainable project management in urban development projects: a case study of the Greater Port Harcourt City Development Project in Rivers State. Sustain Dev Plan VIII 210:209–219

Jagarnath M, Thambiran T, Gebreslasie M (2020) Heat stress risk and vulnerability under climate change in Durban metropolitan, South Africa—identifying urban planning priorities for adaptation. Clim Change 163(2):807–829. https://doi.org/10.1007/s10584-020-02908-x

Jiboye AD (2011) Sustainable urbanization: issues and challenges for effective urban governance in Nigeria. J Sustain Dev 4(6):211–224

Journal du Cameroun (2017) 29 year old lighting up Cameroon through renewable clean energy. https://www.journalducameroun.com/en/29-year-old-lights-cameroon-renewable-clean-energy/. Retrieved 17 Dec 2018

Kalu OH, Ibiam J, Stephen NC, Ijeoma UJ (2014) Rural infrastructure development: a tool for resolving urbanization crisis. A case study of South-Eastern Nigeria. World Environ 4(1):14–21

Kanu I, Achi OK (2011) Industrial effluents and their impact on water quality of receiving rivers in Nigeria. J Appl Technol Environ Sanit 1(1):75–86

Karekezi S, Majoro L (2002) Improving modern energy services for Africa's urban poor. Energy Policy 30(11):1015–1028. https://doi.org/10.1016/S0301-4215(02)00055-1

Keil R (2003) Urban political ecology. Urban Geogr 24(8):723–738. https://doi.org/10.2747/0272-3638.24.8.723

Kenfack J, Fogue M, Hamandjoda O, Tatietse TT (2011) Promoting renewable energy and energy efficiency in Central Africa: Cameroon case study. World renewable energy congress-Sweden; 8–13 May; 2011; Linköping; Sweden, 2602–2609. Retrieved from http://www.ep.liu.se/ecp_article/index.en.aspx?issue=057;vol=10;article=42

Khan H (2017) Top 10 countries with highest deforestation rate in the world 2017. World Blaze. http://www.Worldblaze.In/Countries-With-Highest-Deforestation-Rate-In-The-World/. Retrieved 8 Jan 2018

Kroon FJ, Robinson CJ, Dale AP (2009) Integrating knowledge to inform water quality planning in the Tully-Murray basin, Australia. Mar Freshw Res 60(11):1183. https://doi.org/10.1071/MF08349

Lagosians Lament as flood hits Lekki, Victoria Island (2017) Channels television. https://www.Channelstv.Com/2017/07/08/Lagosians-Lament-As-Flood-Hits-Lekki-Victoria-Island/. Retrieved 30 Nov 2017

Lanrewaju AF (2012) Urbanization, housing quality and environmental degeneration in Nigeria. J Geogr Region Plan 5(16):422–429

Laue F, Adegun OB, Ley A (2022) Heat stress adaptation within informal, low-income urban settlements in Africa. Sustainability 14(13), Article 13. https://doi.org/10.3390/su14138182

Levin S, Xepapadeas T, Crepin A-S, Norberg J, de Zeeuw A, Folke C et al (2013) Social-ecological systems as complex adaptive systems: modeling and policy implications. Environ Dev Econ 18(2):111–132. https://doi.org/10.1017/S1355770X12000460

Levin S, Xepapadeas T, Crepin A-S, Norberg J, de Zeeuw A, Folke C et al (2013) Social-ecological systems as complex adaptive systems: modeling and policy implications. Environ Dev Econ 18(2):111–132. https://doi.org/10.1017/S1355770X12000460

Lewis AO (1977) Nigeria's third national development plan, 1975–80: an appraisal of objectives and policy frame. Dev Econ 15(1):60–79. https://doi.org/10.1111/J.1746-1049.1977.Tb00370.X

Li X, Stringer LC, Dallimer M (2022a) The impacts of urbanisation and climate change on the urban thermal environment in Africa. Climate 10(11), Article 11. https://doi.org/10.3390/cli10110164

Li X, Stringer LC, Dallimer M (2022) The role of blue green infrastructure in the urban thermal environment across seasons and local climate zones in East Africa. Sustain Cities Soc 80:103798. https://doi.org/10.1016/j.scs.2022.103798

Lindley S, Pauleit S, Yeshitela K, Cilliers S, Shackleton C (2018) Rethinking urban green infrastructure and ecosystem services from the perspective of sub-Saharan African cities. Landsc Urban Plan 180:328–338. https://doi.org/10.1016/j.landurbplan.2018.08.016

Lukacs M (2014) New, privatized African City heralds climate apartheid. The Guardian International Edition. https://www.theguardian.com/environment/true-north/2014/jan/21/new-privatized-african-city-heralds-climate-apartheid
Maclean BKR (2018) Biya wins again in Cameroon as crackdown disrupts anglophone vote. The Guardian. Retrieved from https://www.theguardian.com/world/2018/oct/22/paul-biya-cameroon-85-year-old-president-wins-re-election-landslide
Madlener R, Sunak Y (2011) Impacts of urbanization on urban structures and energy demand: What can we learn for urban energy planning and urbanization management? Sustain Cities Soc 1(1):45–53. https://doi.org/10.1016/j.scs.2010.08.006
Magoni M (2017) Energy systems and water and food nexus. In: Colucci A, Magoni M, Menoni S (eds) Peri-urban areas and food-energy-water nexus: sustainability and resilience strategies in the age of climate change, pp 125–129. https://doi.org/10.1007/978-3-319-41022-7_16
Mahabir R, Crooks A, Croitoru A, Agouris P (2016) The study of slums as social and physical constructs: challenges and emerging research opportunities. Reg Stud Reg Sci 3(1):399–419. https://doi.org/10.1080/21681376.2016.1229130
Mahmoud IM, Solan S, Mohammed A, Chapman HM, Dunn A, Odigha O, Laurance WF (2017) Alternative routes for a proposed Nigerian superhighway to limit damage to rare ecosystems and wildlife. Trop Conserv Sci 10:1–10. https://doi.org/10.1177/1940082917709274
Mallett A (2007) Social acceptance of renewable energy innovations: the role of technology cooperation in urban Mexico. Energy Policy 35(5):2790–2798. https://doi.org/10.1016/j.enpol.2006.12.008
Marks D, Connell J (2023) Unequal and unjust: the political ecology of Bangkok's increasing urban heat island. Urban Stud. https://doi.org/10.1177/00420980221140999
Martínez DM, Ebenhack BW (2008) Understanding the role of energy consumption in human development through the use of saturation phenomena. Energy Policy 36(4):1430–1435. https://doi.org/10.1016/j.enpol.2007.12.016
Martinot E (2001) Renewable energy investment by the World Bank. Energy Policy 29(9):689–699. https://doi.org/10.1016/S0301-4215(00)00151-8
Mas'ud AA, Vernyuy Wirba A, Muhammad-Sukki F, Mas'ud IA, Munir AB, Md Yunus N (2015) An assessment of renewable energy readiness in Africa: case study of Nigeria and Cameroon. Renew Sustain Energy Rev 51:775–784. https://doi.org/10.1016/j.rser.2015.06.045
Mbi JN (n.d.) An assessment of solar and bioenergy potential in Cameroon: towards sustainable development. Retrieved from https://www.academia.edu/7351787/An_Assesment_of_Solar_and_Bioenergy_Potential_in_Cameroon_Towards_Sustainable_Development
Mbih RA, Ndzeidze SK, Driever SL, Bamboye GF (2014) The Bamendjin dam and its implications in the Upper Noun Valley, Northwest Cameroon. J Sustain Dev 7(6):123–132. https://doi.org/10.5539/jsd.v7n6p123
Mboumboue E, Njomo D (2016) Potential contribution of renewables to the improvement of living conditions of poor rural households in developing countries: Cameroon's case study. Renew Sustain Energy Rev 61:266–279. https://doi.org/10.1016/j.rser.2016.04.003
Mcdonnell T (2017) Slum dwellers in Lagos, Africa's Biggest Megacity, are now living in Canoes: Goats and Soda: NPR. https://Www.Npr.Org/Sections/Goatsandsoda/2017/05/15/528461093/Slum-Dwellers-In-Africas-Biggest-Megacity-Are-Now-Living-In-Canoes. Retrieved 8 Jan 2018
Mckinney ML (2008) Effects of urbanization on species richness: a review of plants and animals. Urban Ecosyst 11(2):161–176
Mclellan BC, Chapman AJ, Aoki K (2016) Geography, urbanization and lock-in—considerations for sustainable transitions to decentralized energy systems. J Clean Prod 128(C):77–96. https://doi.org/10.1016/j.jclepro.2015.12.092
Mensah CA (2014) Destruction of urban green spaces: a problem beyond urbanization in Kumasi City (Ghana). Am J Environ Prot 3(1), Article 1. https://doi.org/10.11648/j.ajep.20140301.11
Minucci G, Karmaoui A (2017) Exploring the water-food-energy and climate nexus: insights from the Moroccan Draa Valley. In: Colucci A, Magoni M, Menoni S (eds) Peri-urban areas and

food-energy-water nexus: sustainability and resilience strategies in the age of climate change, pp 89–97. https://doi.org/10.1007/978-3-319-41022-7_11
Molua EL (2002) Climate variability, vulnerability and effectiveness of farm-level adaptation options: the challenges and implications for food security in Southwestern Cameroon. Environ Dev Econ 7(3):529–545
Molua EL (2012) Discourse on climate-smart agriculture for REDD+ strategy in the Congo Basin
Molua E, Molua EL (2007) The economic impact of climate change on agriculture in Cameroon (SSRN Scholarly Paper No. ID 1016260). Retrieved from Social Science Research Network website. https://papers.ssrn.com/abstract=1016260
Monforti F, Belward A, Bisselink B, Bódis K, Brink A, Dallemand J et al (2011) Renewable energies in Africa: current knowledge. European Commission, Joint Research Centre (JRC), Ispra
Mueller V, Gray C (2018) Heat and adult health in China. Popul Environ 40(1):1–26. https://doi.org/10.1007/s11111-018-0294-6
Muh E, Amara S, Tabet F (2018) Sustainable energy policies in Cameroon: a holistic overview. Renew Sustain Energy Rev 82:3420–3429. https://doi.org/10.1016/j.rser.2017.10.049
Muhammad F, Wonorahardjo S (2020) Building material in the perspective of energy efficiency and thermal environment in TOD area. IOP Conf Ser Earth Environ Sci 532(1):012021. https://doi.org/10.1088/1755-1315/532/1/012021
Mukherji A, Thorne P, Cheung WWL, Connors SL, Garschagen M, Geden O, Hayward B, Simpson NP, Totin E, Blok K, Eriksen S, Fischer E, Garner G, Guivarch C, Haasnoot M, Hermans T, Ley D, Lewis J, Nicholls Z et al (2023) AR6 Synthesis Report: Climate Change 2023 (IPCC Sixth Assessment Report)
Murphy JT (2001) Making the energy transition in rural east Africa: is leapfrogging an alternative? Technol Forecast Soc Chang 68(2):173–193. https://doi.org/10.1016/S0040-1625(99)00091-8
Musa (2007) Cameroon police kill students in blackout protest. https://uk.reuters.com/article/uk-cameroon-police-students-idUKL12114892007112. Retrieved 22 Nov 2018
Nassary EK, Msomba BH, Masele WE, Ndaki PM, Kahangwa CA (2022) Exploring urban green packages as part of Nature-based Solutions for climate change adaptation measures in rapidly growing cities of the Global South. J Environ Manag 310:114786. https://doi.org/10.1016/j.jenvman.2022.114786
Ncongwane KP, Botai JO, Sivakumar V, Botai CM (2021) A literature review of the impacts of heat stress on human health across Africa. Sustainability, 13(9), Article 9. https://doi.org/10.3390/su13095312
Ndi FA (2017) Land grabbing, local contestation, and the struggle for economic gain: insights from Nguti Village, South West Cameroon. SAGE Open 7(1):2158244016682997. https://doi.org/10.1177/2158244016682997
Ndi FA (2017) Land grabbing, local contestation, and the struggle for economic gain: insights from Nguti Village, South West Cameroon. SAGE Open 7(1):2158244016682997. https://doi.org/10.1177/2158244016682997
Ndi FA, Batterbury S (2017) Land grabbing and the axis of political conflicts: insights from Southwest Cameroon. Afr Spectr 52(1):33–63. https://doi.org/10.1177/000203971705200102
Nfah EM, Ngundam JM (2009) Feasibility of pico-hydro and photovoltaic hybrid power systems for remote villages in Cameroon. Renew Energy 34(6):1445–1450. https://doi.org/10.1016/j.renene.2008.10.019
Nigeria Historical Development of Urban Centers (1991) The library of congress country studies. U.S. Library of Congress. http://countrystudies.us/nigeria/48.htm. Retrieved 9 March 2018
Nigeria's River Basins and Agricultural Development (2009) eWASH, Nigeria's news on water, sanitation and hygiene. http://Assemblyonline.Info/Nigerias-River-Basins-And-Agricultural-Development/. Retrieved 9 March 2018
Njoh AJ (2003) Barriers to community participation in development planning: lessons from the Mutengene (Cameroon) Self-Help Water Project. Community Dev J 37(3):233–248. https://doi.org/10.1093/cdj/37.3.233

Nupe Kingdom (2014) Eneke the bird. https://Enekethebird.Wordpress.Com/Tag/Nupe-Kingdom/. Retrieved 2 March 2018

Nwokoro IIC, Dekolo SO (2012) Land use change and environmental sustainability: the case of Lagos Metropolis. Sustain City VII: Urban Regen Sustain 1:157–167

Nwuba CC, Kalu IU, Umeh JA (2015) Determinants of homeownership affordability in Nigeria's urban housing markets. Int J Hous Mark Anal 8(2):189–206

Nyambod EM (2010) Environmental consequences of rapid urbanisation: Bamenda City, Cameroon. J Environ Prot 01(01):15–23. https://doi.org/10.4236/jep.2010.11003

O'Brien G, Hope A (2010) Localism and energy: negotiating approaches to embedding resilience in energy systems. Energy Policy 38(12):7550–7558. https://doi.org/10.1016/j.enpol.2010.03.033

O'Brien G, O'Keefe P, Rose J (2007) Energy, poverty and governance. Int J Environ Stud 64(5):605–616. https://doi.org/10.1080/00207230600841385

Odjugo PAO et al (2010) General overview of climate change impacts in Nigeria. J Hum Ecol 29(1):47–55

OECD (2012) Energy. OECD, Paris

OECD/SWAC (2020) Africa's urbanisation dynamics 2020: Africapolis, mapping a new urban geography. OECD Publishing. https://read.oecd.ilibrary.org/development/africa-s-urbanisation-dynamics-2020_b6bccb81-en

Offon C (2017) Cross river superhighway: why the threat still stands. Enviro News Nigeria. https://www.environewsnigeria.com/cross-river-superhighway-threat-still-stands/. Retrieved 30 Nov 2017

Ogundele A (2022) Makoko 2.jpg. Wikimedia Commons. https://commons.wikimedia.org/wiki/File:Makoko_2.jpg

Ogundipe S (2017) Special Report: benue flood: two weeks after, it's still pains, anguish and sorrow for displaced victims. Premium Times Nigeria. https://www.Premiumtimesng.Com/News/Headlines/243397-Special-Report-Benue-Flood-Two-Weeks-After-Its-Still-Pains-Anguish-And-Sorrow-For-Displaced-Victims.Html. Retrieved 30 Nov 2017

Ogunlesi T (2017) Inside Makoko: danger and ingenuity in the world's biggest floating slum. The Guardian. https://www.Theguardian.Com/Cities/2016/Feb/23/Makoko-Lagos-Danger-Ingenuity-Floating-Slum. Retrieved 8 Jan 2018

Ogunsi S (2016) The Pre-Colonial Benin empire. The 234 Project. https://the234project.com/history/nigeria/the-pre-colonial-benin-empire/. Retrieved 2 March 2018

Oke TR (1982) The energetic basis of the urban heat island. Q J R Meteorol Soc 108(455):1–24. https://doi.org/10.1002/qj.49710845502

Okhankhuele OT, Opafunso OZ (2013) Causes and consequences of rural-urban migration Nigeria: a case study of Ogun waterside local government area of Ogun State, Nigeria. Br J Arts Soc Sci 16(1):185–194

Okolie A (2017) Benue's devastating flood. Thisdaylive. https://www.thisdaylive.com/index.php/2017/09/10/benues-devastating-flood/. Retrieved 30 Nov 2017

Olujimi J (2009) Evolving a planning strategy for managing urban sprawl in Nigeria. J Hum Ecol 25(3):201–208

Onuoha M (2017) A 5-mile island built to save Lagos's economy has a worrying design flaw. Quartz Africa. https://qz.com/africa/923142/the-flaw-in-the-construction-of-eko-atlantic-island-in-lagos/. Retrieved 30 Nov 2017

Opoko AP, Oluwatayo A, Opoko AP, Oluwatayo A (2014) Trends in urbanisation: implication for planning and low-income housing delivery in Lagos, Nigeria. Archit Res 4(1a):15–26

Orimoloye IR, Mazinyo SP, Kalumba AM, Ekundayo OY, Nel W (2019) Implications of climate variability and change on urban and human health: a review. Cities 91:213–223. https://doi.org/10.1016/j.cities.2019.01.009

Osibanjo O, Daso AP, Gbadebo AM (2011) The impact of industries on surface water quality of River Ona and River Alaro in Oluyole industrial estate, Ibadan, Nigeria. Afr J Biotech 10(4):696–702

Ouedraogo NS (2013) Energy consumption and human development: evidence from a panel cointegration and error correction model. Energy 63:28–41. https://doi.org/10.1016/j.energy.2013.09.067

Oyo Empire (2010) Afropedea. http://www.Afropedea.Org/Oyo-Empire. Retrieved 2 March 2018

Oyeleye Oyewale I (2013) Challenges of urbanization and urban growth in Nigeria. Am J Sustain Cities Soc 1(2):79–95

Oyewole IO, Awolola TS (2006) Impact of urbanisation on bionomics and distribution of malaria vectors in Lagos, Southwestern Nigeria. J Vector Borne Dis 43(4):173–178

Parker (2008) A shock to the ecosystem: neoliberalism and the environment. Well Sharp. https://Wellsharp.Wordpress.Com/2008/07/01/A-Shock-To-The-Ecosystem-Neoliberalism-And-The-Environment/. Retrieved 9 March 2018

Parkes B, Buzan JR, Huber M (2022) Heat stress in Africa under high intensity climate change. Int J Biometeorol 66(8):1531–1545. https://doi.org/10.1007/s00484-022-02295-1

Paun E (2010) The developmental potential of cities and regions in the context of globalization. Geopolit Hist Int Relat 2(1):158–163

Pemunta NV (2018) The logic of benevolent capitalism: the duplicity of Sithe Global Sustainable Oils Cameroon land grab and deforestation scheme as sustainable investment. Int J Glob Environ Issues 17(1):80–109

Pineau P-O (2002) Electricity sector reform in Cameroon: is privatization the solution? Energy Policy 30(11–12):999–1012

Plastow R (2010) Neoliberalism in environmental governance: a paradoxical double movement? University of Exeter. https://www.Academia.Edu/2703516/Neoliberalism_in_Environmental_Governance_A_Paradoxical_Double_Movement. Retrieved Nov 2017

Population Census in Nigeria (2014) Nigerian Observer. http://Nigerianobservernews.Com/2014/12/Population-Census-In-Nigeria/. Retrieved 2 March 2018

Portner H-O et al (2022) Climate change 2022: impacts, adaptation and vulnerability. Intergovernmental Panel on Climate Change. https://report.ipcc.ch/ar6wg2/pdf/IPCC_AR6_WGII_FinalDraft_FullReport.pdf

Potsdam Institute for Climate Impact Research and Climate (2013) Turn down the heat: climate extremes, regional impacts, and the case for resilience, p 252. World Bank. https://openknowledge.worldbank.org/handle/10986/14000

Reddy S, Painuly JP (2004) Diffusion of renewable energy technologies—barriers and stakeholders' perspectives. Renew Energy 29(9):1431–1447. https://doi.org/10.1016/j.renene.2003.12.003

Ringard J, Dieppois B, Rome S, Diedhiou A, Pellarin T, Konaré A, Diawara A, Konaté D, Dje BK, Katiellou GL, Seidou Sanda I, Hassane B, Vischel T, Garuma GF, Mengistu G, Camara M, Diongue A, Gaye AT, Descroix L (2016) The intensification of thermal extremes in west Africa. Global Planet Change 139:66–77. https://doi.org/10.1016/j.gloplacha.2015.12.009

Rockstrom J, Steffen W, Noone K, Persson A, Chapin F, Lambin E et al (2009) Planetary boundaries: exploring the safe operating space for humanity. Ecol Soc 14(2)

Rockström J, Steffen W, Noone K, Persson Å, Chapin FS, Lambin EF, Foley JA (2009) A safe operating space for humanity. Nature 461(7263):472–475. https://doi.org/10.1038/461472a

Roser M, Ritchie H (2013) Indoor air pollution. Our world in data. Retrieved from https://ourworldindata.org/indoor-air-pollution

Roser M, Rodés-Guirao L (2013) Future population growth. Our world in data. https://ourworldindata.org/future-population-growth

Roy P (2011) Pursuing urban political ecology for a sustainable and just urban environment. Indian J Spat Sci II(1), Article 3-7

Rutter P, Keirstead J (2012) A brief history and the possible future of urban energy systems. Energy Policy 50:72–80. https://doi.org/10.1016/j.enpol.2012.03.072

Sagar AD (2005) Alleviating energy poverty for the world's poor. Energy Policy 33(11):1367–1372. https://doi.org/10.1016/j.enpol.2004.01.001

Salami RO, von Meding J, Giggins H (2018) The urban political ecology of flood vulnerability in the core area of Ibadan metropolis, Nigeria. In: Natural hazards and disaster risk reduction policies. Il Sileno Edizioni, pp 36–50. https://nova.newcastle.edu.au/vital/access/manager/Repository/uon:35714;jsessionid=9F022B6A1EB1ED2C042C5B8F6040EABA?view=grid&f0=sm_identifier%3A%22http%3A%2F%2Fhdl.handle.net%2F1959.13%2F1407290%22&sort=ss_dateNormalized+asc%2Csort_ss_title+asc

Sally Z, Gaskin SJ, Folifac F, Kometa SS (2014) The effect of urbanization on community-managed water supply: case study of Buea, Cameroon. Community Dev J 49(4):524–540. https://doi.org/10.1093/cdj/bst054

Schlosberg D (2007) Defining environmental justice theories, movements, and nature. Oxford University Press, New York, NY

Serdeczny O, Adams S, Baarsch F, Coumou D, Robinson A, Hare W, Schaeffer M, Perrette M, Reinhardt J (2017) Climate change impacts in Sub-Saharan Africa: from physical changes to their social repercussions. Reg Environ Change 17(6):1585–1600. https://doi.org/10.1007/s10113-015-0910-2

Shehu J (2017) Urbanization in the 20th century: some reflections on Katsina Town, Nigeria. J Geogr Assoc Tanzania 36(1):1–11

Silver J (2015) Disrupted Infrastructures: an urban political ecology of interrupted electricity in Accra. Int J Urban Reg Res 39(5):984–1003. https://doi.org/10.1111/1468-2427.12317

Smit W, Parnell S (2012) Urban sustainability and human health: an African perspective. Curr Opin Environ Sustain 4(4):443–450. https://doi.org/10.1016/j.cosust.2012.07.004

Smith JW, Leahy JE, Anderson DH, Davenport MA (2013) Community/agency trust and public involvement in resource planning. Soc Nat Resour 26(4):452–471. https://doi.org/10.1080/08941920.2012.678465

Sokoto Caliphate (2018) Harvard Divinty School, Religious Literacy Poject. https://rlp.hds.harvard.edu/faq/sokoto-caliphate. Retrieved Sept 2018

Solargis (2014). https://www.get-invest.eu/market-information/cameroon/renewable-energy-potential/. Retrieved 3 Apr 2019

Sorlini S, Palazzini D, Sieliechi JM, Ngassoum MB (2013) Assessment of physical-chemical drinking water quality in the Logone Valley (Chad-Cameroon). Sustainability 5(7):3060–3076. https://doi.org/10.3390/su5073060

Sovacool BK (2012) The political economy of energy poverty: a review of key challenges. Energy Sustain Dev 16(3):272–282

Spinoni J, Barbosa P, Füssel H, McCormick N, Vogt JV, Dosio A (2021) Global population-weighted degree-day projections for a combination of climate and socio-economic scenarios. Int J Climatol 41(11):5447–5464. https://doi.org/10.1002/joc.7328

Steer A, Hanson C (2015) Biofuels are not a green alternative to fossil fuels. The Guardian. Retrieved from http://www.theguardian.com/environment/2015/jan/29/biofuels-are-not-the-green-alternative-to-fossil-fuels-they-are-sold-as

Stirling A (2014) Transforming power: social science and the politics of energy choices. Energy Res Soc Sci 1:83–95. https://doi.org/10.1016/j.erss.2014.02.001

Sulaiman C, Abdul-Rahim AS, Chin L, Mohd-Shahwahid HO (2017) Wood fuel consumption and mortality rates in Sub-Saharan Africa: evidence from a dynamic panel study. Chemosphere 177:224–231. https://doi.org/10.1016/j.chemosphere.2017.03.019

Sun Y, Asante F, Birner R (2010) Opportunities and challenges of community-based rural drinking water supplies: an analysis of water and sanitation committees in Ghana. IDEAS Working Paper Series from RePEc. Retrieved from http://search.proquest.com/docview/1698099609/?pq-origsite=primo

Swyngedouw E (2006) Circulations and metabolisms: (hybrid) natures and (Cyborg) cities. Sci Cult 15(2):105–121. https://doi.org/10.1080/09505430600707970

Swyngedouw E, Kaika M (2014) Urban political ecology. Great promises, deadlock… And new beginnings? Documents D'anàlisi Geogràfica 60(3):459–481

Taiwo OJ, Olaniran HD, Osayomi T (2012) Perceived causes, exposures and adjustments to seasonal heat in different residential areas in Ibadan, Nigeria. Environmentalist 32(4):405–414. https://doi.org/10.1007/s10669-012-9403-8

Takebayashi H, Misaka I, Akagawa H (2020) Chapter 2—Adaptation measures and their performance. In: Takebayashi H, Moriyama M (eds) Adaptation measures for urban heat islands. Academic Press, London, pp 9–37. https://doi.org/10.1016/B978-0-12-817624-5.00002-6

Tamba JG (2017) An analysis of Cameroon's energetic system, 2001–2010. Energy Sources Part B 12(3):216–222. https://doi.org/10.1080/15567249.2015.1060547

Taylor B, de Loë RC (2012) Conceptualizations of local knowledge in collaborative environmental governance. Geoforum 43(6):1207–1217. https://doi.org/10.1016/j.geoforum.2012.03.007

Tchouaha S (2012) Hydropower in Cameroon. Bachelors Dissertation. University of Gavle, Sweden

The Brundtland Report 'Our Common Future' (1987). https://www.sustainabledevelopment2015.org/AdvocacyToolkit/index.php/earth-summit-history/historical-documents/92-our-common-future. Retrieved 25 Oct 2018

The Commonwealth (2019) Cameroon History. http://thecommonwealth.org/our-member-countries/cameroon/history. Retrieved 10 Feb 2019

The World Bank (2023) Urban population (% of total population)—Sub-Saharan Africa | Data. The World Bank Data. https://data.worldbank.org/indicator/SP.URB.TOTL.IN.ZS?end=2021&locations=ZG&name_desc=false&start=1960

The World Bank Group (2016) Independent evaluation group (IEG). http://Lnweb90.Worldbank.Org/Oed/Oeddoclib.Nsf/Docunidviewforjavasearch/Fe7ba13642e3e0d7852567f5005d85cf. Retrieved 9 March 2018

Thorn JPR, Biancardi Aleu R, Wijesinghe A, Mdongwe M, Marchant RA, Shackleton S (2021) Mainstreaming nature-based solutions for climate resilient infrastructure in peri-urban sub-Saharan Africa. Landsc Urban Plan 216:104235. https://doi.org/10.1016/j.landurbplan.2021.104235

Titz A, Chiotha SS (2019) Pathways for sustainable and inclusive cities in Southern and Eastern Africa through urban green infrastructure? Sustainability 11(10), Article 10. https://doi.org/10.3390/su11102729

Transparency International—Cameroon (2018). https://www.transparency.org/country/CMR. Retrieved 7 Feb 2019

Tuholske C, Caylor K, Funk C, Verdin A, Sweeney S, Grace K, Peterson P, Evans T (2021) Global urban population exposure to extreme heat. Proc Natl Acad Sci 118(41):e2024792118. https://doi.org/10.1073/pnas.2024792118

Ukpong C (2017) Nigerian government endorses cross river's 260km super highway. Premium Times Nigeria. https://www.premiumtimesng.com/regional/south-south-regional/232831-nigerian-govt-endorses-cross-rivers-260km-super-highway.html. Retrieved 8 Jan 2018

Unaegbu E (2017) Cross river super highway: a development without the people. Businessday. https://Www.Businessdayonline.Com/Cross-River-Super-Highway-Development-Without-People/. Retrieved 8 Jan 2018

UNDP (2009) The energy access situation in developing countries. World Health Organization. https://www.undp.org/content/dam/undp/library/Environment%20and%20Energy/Sustainable%20Energy/energy-access-situation-in-developing-countries.pdf. Retrieved 16 Jan 2019

UNDP (2016) Delivering sustainable energy in a changing climate: strategy note on sustainable energy, 2017–2021

Unemployment/Underemployment Report Q4 (2016) Nigeria data portal. http://Nigeria.Opendataforafrica.Org/Nglbfrstat2016/Unemployment-Underemployment-Report-Q4-2016. Retrieved 8 Jan 2018

UNFCCC (2001) CDM: about CDM. (n.d.). https://cdm.unfccc.int/about/index.html. Retrieved 22 Jan 2019

UniAfrica (2017) Cameroon: Italian Enerray refines 30 MW solar project in Garoua, North Region. https://www.cac-international.com/cameroon-italian-enerray-refines-30-mw-solar-project-in-garoua-north-region/?lang=en. Retrieved 17 Dec 2018

United Nations Development Program (UNDP) Human Development Reports. (2018). http://hdr.undp.org/en/countries/profiles/CMR. Retrieved 23 Jan 2019

UN Habitat (2014) Cameroon Impact Story – UN-Habitat. https://unhabitat.org/cameroon-impact-story/. Retrieved 19 March 2019

United Nations (2015) Sustainable development knowledge platform. https://sustainabledevelopment.un.org/sdg7. Retrieved 25 Oct 2018

United Nations (2018) 68% of the world population projected to live in urban areas by 2050, says UN. UN DESA | United Nations Department of Economic and Social Affairs website: https://www.un.org/development/desa/en/news/population/2018-revision-of-world-urbanization-prospects.html. Retrieved 8 Nov 2018

United Nations Human Settlements Programme (Un-Habitat) (2017) Un-habitat. https://Unhabitat.Org/. Retrieved 8 Jan 2018

United Nations Population Division (2023) Urban population (% of total population)—Sub-Saharan Africa | Data. Urban population (% of total population)—Sub-Saharan Africa. https://data.worldbank.org/indicator/SP.URB.TOTL.IN.ZS?locations=ZG

US Department of Commerce N (n.d.) Wet bulb globe temperature. NOAA's National Weather Service. https://www.weather.gov/car/WBGT. Retrieved 10 Feb 2023

Van de Walle J, Brousse O, Arnalsteen L, Brimicombe C, Byarugaba D, Demuzere M, Jjemba E, Lwasa S, Misiani H, Nsangi G, Soetewey F, Sseviiri H, Thiery W, Vanhaeren R, Zaitchik BF, van Lipzig NPM (2022) Lack of vegetation exacerbates exposure to dangerous heat in dense settlements in a tropical African city. Environ Res Lett 17(2):024004. https://doi.org/10.1088/1748-9326/ac47c3

Voskamp IM, de Luca C, Polo-Ballinas MB, Hulsman H, Brolsma R (2021) Nature-based solutions tools for planning urban climate adaptation: state of the art. Sustainability 13(11), Article 11. https://doi.org/10.3390/su13116381

Walker G (2009) Beyond distribution and proximity: exploring the multiple spatialities of environmental justice. Antipode 41(4):614–636

Walker RV, Poponi D, Lefevre B (n.d.) Advancing toward a more sustainable urban energy system 10

Welborn L (2018) Africa and climate change: projecting vulnerability and adaptive capacity (Africa Report, p. 24). Institute for Security Studies. https://issafrica.s3.amazonaws.com/site/uploads/ar14.pdf

Wijesinghe A, Thorn JPR (2021) Governance of urban green infrastructure in informal settlements of Windhoek, Namibia. Sustainability 13(16), Article 16. https://doi.org/10.3390/su13168937

Wilby RL, Kasei R, Gough KV, Amankwaa EF, Abarike M, Anderson NJ, Codjoe SNA, Griffiths P, Kaba C, Abdullah K, Kayaga S, Matthews T, Mensah P, Murphy C, Yankson PWK (2021) Monitoring and moderating extreme indoor temperatures in low-income urban communities. Environ Res Lett 16(2):024033. https://doi.org/10.1088/1748-9326/abdbf2

Williams KN, Northcross AL, Graham JP (2015) Health impacts of household energy use: indicators of exposure to air pollution and other risks. Bull World Health Organ 93(7):507–508. https://doi.org/10.2471/BLT.14.144923

Winsor M (2015) Nigeria's new city: Eko Atlantic construction in Lagos fuels criticism and praise. International Business Times. https://www.ibtimes.com/nigerias-new-city-eko-atlantic-construction-lagos-fuels-criticism-praise-2048964. Retrieved 8 Jan 2018

Wirba AV, Abubakar Mas'ud A, Muhammad-Sukki F, Ahmad S, Mat Tahar R, Abdul Rahim R et al (2015) Renewable energy potentials in Cameroon: prospects and challenges. Renew Energy 76:560–565. https://doi.org/10.1016/j.renene.2014.11.083

Wolch JR, Byrne J, Newell JP (2014) Urban green space, public health, and environmental justice: the challenge of making cities 'just green enough.' Landsc Urban Plan 125:234–244. https://doi.org/10.1016/j.landurbplan.2014.01.017

Worku H (2017) Integrating climate change adaptation strategies in urban planning and landscape design of Addis Ababa City, Ethiopia. Environ Qual Manag 27(1):5–21. https://doi.org/10.1002/tqem.21514

World and Regional Statistics, National Data, Maps, Rankings (2016) Knoema. https://knoema.com/atlas/Nigeria/topics/Water/Sanitation-Rural-Population/Proportion-of-rural-population-served-with-improved-sanitation. Retrieved 8 Jan 2018

World Bank (2015) The Lom Pangar hydropower dam is set to power Cameroon. World Bank website. http://www.worldbank.org/en/news/video/2015/12/04/the-lom-pangar-hydropower-dam-is-set-to-power-cameroon. Retrieved 6 Nov 2017

World Bank (2017) Urban population (% of total) | Data. (n.d.). https://data.worldbank.org/indicator/SP.URB.TOTL.IN.ZS. Retrieved 8 Nov 2018

World Bank Cameroon Overview (2016) http://www.worldbank.org/en/country/cameroon/overview. Retrieved 25 Nov 2017

World Bank Data (2016). https://data.worldbank.org/indicator/EG.ELC.ACCS.ZS?locations=CM. Retrieved 26 Oct 2018

World Bank Open Data (2016) The World Bank. https://data.worldbank.org/indicator/EG.ELC.ACCS.UR.ZS?locations=NG. Retrieved 10 Jan 2018

World Economic Forum (2018) African cities will double in population by 2050. Here are 4 ways to make sure they thrive. (n.d.). World Economic Forum website. https://www.weforum.org/agenda/2018/06/Africa-urbanization-cities-double-population-2050-4%20ways-thrive/. Retrieved 7 Feb 2019

World Population by Country (2017) World population review. http://worldpopulationreview.com/countries/nigeria-population/. Retrieved 8 Jan 2018

Wright CY, Dominick F, Kapwata T, Bidassey-Manilal S, Engelbrecht JC, Stich H, Mathee A, Matooane M (2019) Socio-economic, infrastructural and health-related risk factors associated with adverse heat-health effects reportedly experienced during hot weather in South Africa. Pan Afr Med J 34(40), Article 40. https://doi.org/10.11604/pamj.2019.34.40.17569

Wüstenhagen R, Wolsink M, Bürer MJ (2007) Social acceptance of renewable energy innovation: an introduction to the concept. Energy Policy 35(5):2683–2691. https://doi.org/10.1016/j.enpol.2006.12.001

Yli-Pelkonen V, Kohl J (2005) The role of local ecological knowledge in sustainable urban planning: perspectives from Finland. Sustain Sci Pract Policy 1(1):3–14. https://doi.org/10.1080/15487733.2005.11907960

Yow DM (2007) Urban heat islands: observations, impacts, and adaptation. Geogr Compass 1(6):1227–1251. https://doi.org/10.1111/j.1749-8198.2007.00063.x

Zhang L, Cao H, Han R (2021) Residents' preferences and perceptions toward green open spaces in an urban area. Sustainability 13(3), Article 3. https://doi.org/10.3390/su13031558

Zimmer A (2010) Urban political ecology theoretical concepts, challenges, and suggested future directions. Erdkunde 64(4):343–364. https://doi.org/10.3112/erdkunde.2010.04.04

MIX
Papier aus verantwortungsvollen Quellen
Paper from responsible sources
FSC® C105338
FSC
www.fsc.org

If you have any concerns about our products,
you can contact us on
ProductSafety@springernature.com

In case Publisher is established outside the EU,
the EU authorized representative is:
Springer Nature Customer Service Center GmbH
Europaplatz 3, 69115 Heidelberg, Germany

Printed by Libri Plureos GmbH
in Hamburg, Germany